PAIN MANAGEMENT - RESEARCH AND TECHNOLOGY

BEHAVIORAL STUDY OF AGONIST-EVOKED ACTIVATION OF TRANSIENT RECEPTOR POTENTIAL CHANNELS

PAIN MANAGEMENT - RESEARCH AND TECHNOLOGY

Additional books and e-books in this series can be found on Nova's website under the Series tab.

PAIN MANAGEMENT - RESEARCH AND TECHNOLOGY

BEHAVIORAL STUDY OF AGONIST-EVOKED ACTIVATION OF TRANSIENT RECEPTOR POTENTIAL CHANNELS

MERAB G. TSAGARELI

nova science publishers
New York

Copyright © 2019 by Nova Science Publishers, Inc.

All rights reserved. No part of this book may be reproduced, stored in a retrieval system or transmitted in any form or by any means: electronic, electrostatic, magnetic, tape, mechanical photocopying, recording or otherwise without the written permission of the Publisher.

We have partnered with Copyright Clearance Center to make it easy for you to obtain permissions to reuse content from this publication. Simply navigate to this publication's page on Nova's website and locate the "Get Permission" button below the title description. This button is linked directly to the title's permission page on copyright.com. Alternatively, you can visit copyright.com and search by title, ISBN, or ISSN.

For further questions about using the service on copyright.com, please contact:
Copyright Clearance Center
Phone: +1-(978) 750-8400   Fax: +1-(978) 750-4470   E-mail: info@copyright.com.

NOTICE TO THE READER

The Publisher has taken reasonable care in the preparation of this book, but makes no expressed or implied warranty of any kind and assumes no responsibility for any errors or omissions. No liability is assumed for incidental or consequential damages in connection with or arising out of information contained in this book. The Publisher shall not be liable for any special, consequential, or exemplary damages resulting, in whole or in part, from the readers' use of, or reliance upon, this material. Any parts of this book based on government reports are so indicated and copyright is claimed for those parts to the extent applicable to compilations of such works.

Independent verification should be sought for any data, advice or recommendations contained in this book. In addition, no responsibility is assumed by the publisher for any injury and/or damage to persons or property arising from any methods, products, instructions, ideas or otherwise contained in this publication.

This publication is designed to provide accurate and authoritative information with regard to the subject matter covered herein. It is sold with the clear understanding that the Publisher is not engaged in rendering legal or any other professional services. If legal or any other expert assistance is required, the services of a competent person should be sought. FROM A DECLARATION OF PARTICIPANTS JOINTLY ADOPTED BY A COMMITTEE OF THE AMERICAN BAR ASSOCIATION AND A COMMITTEE OF PUBLISHERS.

Additional color graphics may be available in the e-book version of this book.

Library of Congress Cataloging-in-Publication Data

ISBN: 978-1-53616-501-2
Library of Congress Control Number:2019951529

Published by Nova Science Publishers, Inc. † New York

Contents

Preface		vii
Acknowledgments		ix
Abbreviations		xi
Chapter 1	Introduction	1
Chapter 2	Phylogeny of TRP Channels	5
Chapter 3	Structural-Functional Characteristics of TRP Channels	9
Chapter 4	A Short Review of TRP Channels Superfamily	17
Chapter 5	Thermo-TRP Channels	65
Chapter 6	TRP Channels in Physiological Nociception and Pain	79
Chapter 7	TRP Channels as Therapeutic and Analgesic Targets	113
Chapter 8	TRP Channels and NSAIDs	119
Chapter 9	Concluding Remarks and Perspectives	137

References	**139**
Index	**177**
Related Nova Publications	**183**

PREFACE

Transient receptor potential channels are a superfamily of trans-membrane cation channels involved in transduction in response to a plenty of chemical and physical stimuli. Comprised of four subunits with six trans-membrane helices, these channels can homo- or hetero-tetramerize to create a pore for cation permeation. These channels are found in the plasma membrane and can gate several types of mono- and divalent cations, through the pore following exposure to a stimulus. Transient receptor potential channels have also been implicated as sensors of many physiological and pathological processes including temperature and mechanical sensations, cancers, pain and itch. Almost for last two decades this superfamily of channels has received a tremendous interest because members of this family are involved in a plethora of cell functions and have been identified as causal for many hereditary and acquired diseases.

This book provides up-to-date information on the molecular and functional properties transient receptor potential channels and of my laboratory evidences describing properties of a single channel or portray more general principles. I hope that this book will guide a large reader community through the fascinating world of the transient receptor potential channels superfamily from basic science to pathophysiology and clinic.

In conclusion, I will gratefully accept any comments and notes of my readers.

ACKNOWLEDGMENTS

Almost fifteen years ago, I started collaboration with Earl E. Carstens and Mirela Iodi Carstens from University of California at Davis on the Civilian Research and Development Foundation (CRDF) grant to study the role of transient receptor potential channels in pain and temperature sensation. I wish to express my sincere appreciation to them for introducing me to this new and fascinating area of modern neurobiology. Really, it was a start of new era for my laboratory research.

The book offered to the reader is based on the results of the mentioned grant as well as some grants received from National Science Foundation of Georgia. I am truly grateful to Amanda H. Klein from the lab of Carstens, now working at University of Minnesota, Duluth, and also my lab fellows Lia Nozadze, Nana Tsiklauri and Guliko Gurtskaia for their professional assistance.

I most grateful to my Italian colleagues Pierangelo Geppetti and Romina Nassini from Florence University for fruitful collaboration, and for our nice article that has recently published.

I wish also to express my gratitude to Nova Science Publishers for their generosity and rapidly with which they have made publication of this book.

ABBREVIATIONS

AA (aa)	amino acids
ADP	adenosine diphosphate
AITC	allyl isothiocyanate
AMP	adenosine mono-phosphate
ASA	acetylsalicylic acid
ATP	adenosine triphosphate
CA	cinnamon aldehyde
CGRP	calcitonin gene-related peptide
CHO	Chinese hamster ovary cells
CNS	central nervous system
CeA	central nucleus of amygdala
COPD	chronic obstructive pulmonary disease
COX1	cyclo-oxygenase 1
COX2	cyclo-oxygenase 2
Cryo-EM	cryo-electron microscopy
CSNB	congenital stationary night blindness
DA	diacylglycerol
dsRNA	double-stranded RNA
DRG	dorsal root ganglion
ECM	extracellular matrix
ER	endoplasmic reticulum
FPP	farnesyl pyrophosphate

FRET	fluorescence resonance energy transfer
GPCR	G-protein coupled receptor
GFP	green fluorescent protein
GSC	glioblastoma stem/progenitor-like cell
GWAS	genome-wide association studies
IAG	ibuprofen-acyl glucuronide
IP_3	inositol triphosphate
KO	knockout mice
LELs	late endosomes and lysosomes
LPL	lipoprotein lipase
LSD	lysosomal storage disorder
MOE	main olfactory epithelium
MscL	mechanosensitive large conductance ion channel
mGluR	metabotropic glutamate receptor
MRCA	most recent common ancestor
MLIV	mucolipidosis type IV
NCS-1	neuronal calcium sensor-1
nAChR	neuronal acetylcholine (nicotinic) receptor
NSAIDs	non-steroidal anti-inflammatory drugs
NFaT	nuclear factors in activated T cells
NHERF1	Na^+/H^+ exchanger regulatory factor 1,
NHERF2	and factor 2
NGF	nerve growth factor
NG	nodose ganglion
OSN	olfactory sensory neuron
PDB	protein data bank
PGs	prostaglandins
PIP2	phosphatidyl-inositol biphosphate
PKA	protein kinase A
PLC	phospholipase C
PMCA	plasma membrane calcium ATP
PMD	polycystin-mucolipin domain
POA	preoptic area of hypothalamus
PR	paraneoplastic retinopathy

PSA	prostate-specific antigen
ROMK	renal outer medullary potassium channel
ROS	reactive oxygen species
RT-PCR	reverse transcription polymerase chain reaction
RTX	resiniferatoxin
SFs	selectivity filters of channels
SNI	spared nerve injury
SNP	single-nucleotide polymorphism
SOC	store-operated channel
STIM1	stromal interaction molecule 1
TG	trigeminal ganglion
TGF-β1	transforming growth factor beta 1
TIRF	total internal reflection fluorescence microscopy
TM	trans-membrane domain
TOP	tetragonal opening for polycystins domain
TPPO	tri-phenyl-phosphine oxide
TREK	TWIK-related potassium channel
Trk	tyrosine receptor kinase
Vc	trigeminal subnucleus caudalis
VSMC	vascular smooth muscle cells
WSN	warm-sensitive neurons
WDR	wide dynamic range (neurons)
WT	wild type (mice, rats)

Chapter 1

INTRODUCTION

The Transient Receptor Potential (TRP) channel superfamily is comprised of a large group of cation-permeable channels, which display an extraordinary diversity of roles in sensory signaling and are involved in plethora of animal behaviors. These channels are activated through a wide variety of mechanisms and participate in virtually every sensory modality. In particular, TRPs are critical for sensing the external environment, functioning in vision, thermosensation, olfaction, taste, mechano- and hygro-sensations. Consequently, these channels have a profound impact on animal behavior which includes survival mechanisms in a challenging of environment (Fowler, Montell, 2012).

According to the PubMed data, the number of publications on this topic has risen explosively for last three decades. So far, more than 17.000 publications and 2.500 review articles have been published about TRPs channels (Figure 1).

TRP channels were initially discovered in a blind strain of *Drosophila melanogaster* (Montell, Rubin, 1989). When exposed to prolonged intense light, these spontaneously mutant fruit flies showed transient calcium influx into their photoreceptor cells; this is why the mutant gene was termed *trp*, 'transient receptor potential'. This seminal finding paved the way to the discovery of the first mammalian TRP channels subfamily, called 'canonical' (TRPC1) due to their homology to the *Drosophila* channel. Although the exact function of TRPC1 is still elusive, TRP channels now represent an extended family of 28 members, fulfilling multiple roles in the living organism.

As a general rule, TRP channels are 'cellular sensors' that respond to changes in the cellular environment, including temperature, stretch/pressure, chemicals, oxidation/reduction, osmolarity and pH, both acidic and alkaline. Of note, a number of TRP channels are also activated by natural products, including herbs, spices, venoms and toxins (Kaneko, Szallasi, 2014).

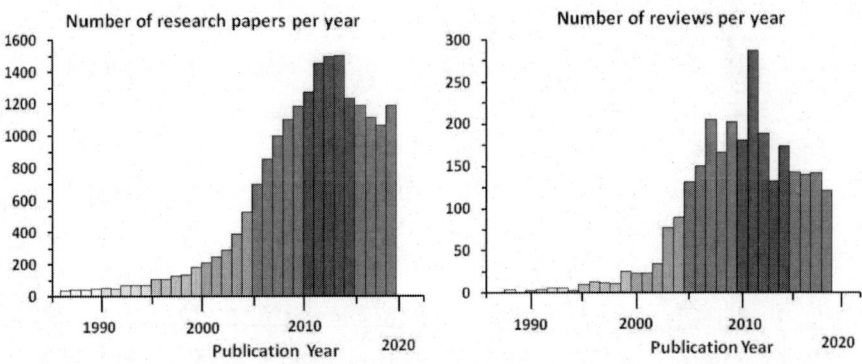

Figure 1. Number of publications in the TRP field in accordance with the PubMed data (March 25, 2019).

The importance of these channels is underscored by the number of genetic diseases caused when they are mutated: skeletal, skin, sensory, ocular, cardiac, and neuronal disturbances all arise from aberrant TRP function. Significant interest in the therapeutic utility of TRP channels persists across disease areas despite these challenges because of the compelling biology (Moran, 2018). Modulating TRP channel activity provides an important way to regulate membrane excitability and intracellular calcium levels. This is reflected by the fact that small molecule compounds modulating different TRP have all entered clinical trials for a variety of diseases. The role of TRPs will be further elucidated in complex diseases of the nervous, intestinal, renal, urogenital, respiratory, and cardiovascular systems in diverse therapeutic areas including pain and itch, pulmonary function, oncology, neurology, visceral organs and genetic diseases.

Numerous experiments have been conducted to investigate the role of TRP channels in sensory signal transductions. These studies presented evidence showing that the peripheral nervous system in particular transduces cutaneous sensory stimuli into electrical signals and transmits them to the central nervous system (CNS). In general, mammalian TRP channels comprise six related protein subfamilies TRP-canonical (TRPC), TRP-vanilloid (TRPV), TRP-melastatin (TRPM), TRP-ankyrin (TRPA), TRP-mucolipin (TRPML), and TRP-polycystic (TRPP). Although the physiological functions of most TRP channels are not well known, their wide distribution indicates that the biological functions and activation mechanisms for these channels are diverse and important. For instance, TRP channels are best recognized for their contributions to sensory transduction, response to temperature, nociceptive stimuli, touch, itch, osmolarity, pheromones or odorants, and other stimuli from both within and outside the cell (Flockerzi, Nilius, 2014; Julius, 2013; Moran, Szallasi, 2018; Tsagareli, 2013; Tsagareli et al., 2019).

Chapter 2

PHYLOGENY OF TRP CHANNELS

The TRP channels are ancient, evolutionarily conserved proteins that function in a wide range of metazoan organisms. These include invertebrates, such as worms, arachnids (e.g., ticks), insects (e.g., flies, mosquitoes and bees) and vertebrates such as zebrafish, mice, and humans. However, the range of TRP channels in some ancient organisms, such as protozoa, is limited to TRPP and TRPML channels (Venkatachalam et al., 2014).

In insects the total number of TRP family members is 13-14, approximately half that of mammalian TRP family members. As shown for mammalian TRP channels, this may suggest that single TRP channels are responsible for integrating diverse sensory inputs to maintain the insect sensory systems comparable to that of mammals. These results demonstrate that there have been both evolutionary conservation and changes in insect TRP channels. In particular, the evolutionary processes accelerated in the TRPA subfamily, indicating divergence in the mechanisms that insects use to detect environmental temperatures (Matsuura et al., 2009).

At least five of the six subfamilies include members that are conserved in animals as divergent as the roundworm *Caenorhabditis elegans*, the fruit fly *Drosophila*, and humans. Representative members of most of the TRP subfamilies have been expressed *in vitro*, and each appears to be a non-selective cation channel. Nevertheless, the modes by which the various TRPs are activated appear to be quite diverse (Latorre et al., 2009; Montell, 2001).

Today, almost 30 years after the first TRP channel was obtained, we can list more than hundred channels within this superfamily (invertebrate and vertebrate combined) organized by their sequence homology into seven subfamilies. There are 33 TRP channel genes in mammals, nearly 60 in zebrafish, 30 in sea squirts, 24 in nematodes, 16 in fruit flies and 1 in yeast. They are subdivided into seven subfamilies on the basis of sequence similarity (Christensen, Corey, 2007) (Figure 2).

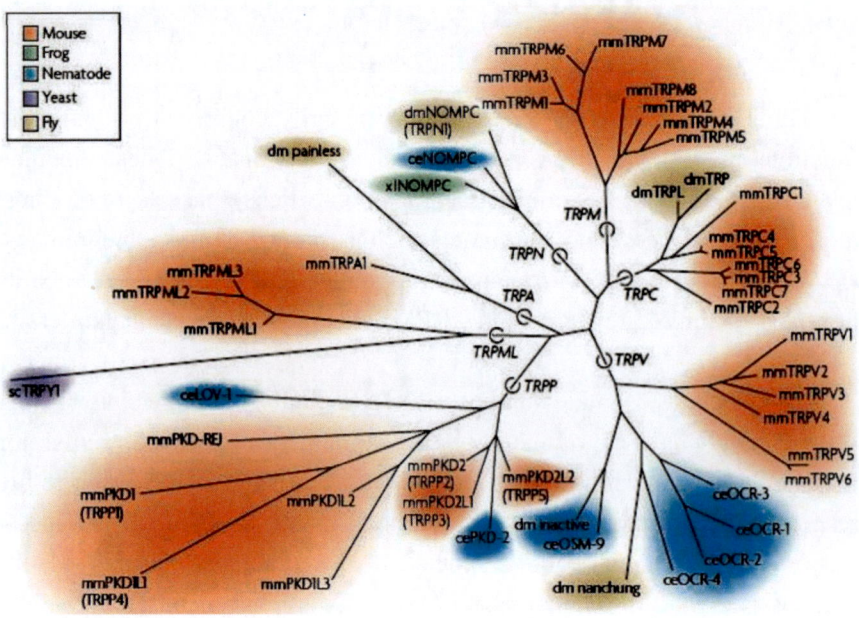

Figure 2. Phylogeny of representative TRP channels. A phylogenic tree was generated in ClustalX by aligning the transmembrane domains of all 33 TRP channels from mouse and some from other species. The seven main branches are denoted by circles at the branch roots. The letters and numbers following TRP denote TRP subfamily and member, respectively. Different species are indicated by colours and by prefixes. ce, *Caenorhabditis elegans*; dm, *Drosophila melanogaster*; mm, *Mus musculus*; sc, *Saccharomyces cerevisiae*; xl, *Xenopus laevis*. (Reproduced from Christensen, Corey, 2007).

All seven subfamilies are represented in flies and worms, although these organisms have fewer TRP channels than humans. Nevertheless, the TRPA subfamily is the largest in insects, even though only one representative exists in mice and humans. This notable difference might reflect a particularly important role for TRPA channels in sensing environmental chemicals and changes in temperature, since poikilothermic animals such as insects are particularly sensitive to heat and cold, and are subjected to a very complex repertoire of compounds in their surroundings (Venkatachalam et al., 2014).

Combined molecular and physiological analyses of nociceptive transduction have revealed important mechanisms in a few species. Particularly, the TRP channels have demonstrated their importance for nociceptive transduction in both vertebrates and invertebrates (Julius, 2013; Walters, 2009). The repertoires of temperature-gated TRP (or thermo-TRP) homologous have changed through vertebrate evolution. Since thermo-TRPs are involved in thermosensation as well as other kinds of sensory detection, variability of repertoires may be associated with adaptation of the organisms to their respective habitat environments (Dhaka et al., 2006; Saito, Shingai, 2006). Distinct members of the thermo-TRP families are present in the invertebrates, *C. elegans* and *D. melanogaster* as illustrated in Figure 3.

Reconstruction of thermo-TRPs history revealed several gene duplications preceding high divergences of protein sequences. Almost all TRPs, including the hot and warm (cold) nociceptors (TRPV1 and TRPA1, respectively), are found across vertebrates and originated in their most recent common ancestor (MRCA). The mild-cold and mild-heat receptors in mammals, TRPM8 and TRPV3, respectively, emerged later at latest in the common ancestor of tetrapod and teleost fishes and became the main components of temperature sensation in mammals. Notably, TRPA1 is a heat receptor in reptiles and TRPV3 is a cold receptor in frogs. This suggests that thermo-receptors underwent a diversification phase in Tetrapoda, probably associated with the adaptation to non-aqueous ecosystems subjected to greater variations in temperature (Blanquart et al., 2019).

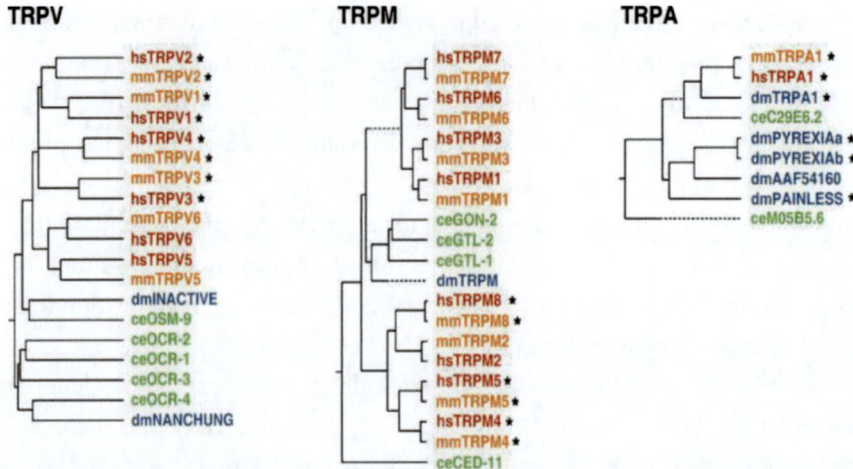

Figure 3. Phylogenies are shown for three families of TRP cation channels: TRPV, TRPM, and TRPA. *H. sapiens*, *M. musculus*, *D. melanogaster*, and *C. elegans* members of these families are color coded. Channels implicated in thermosensation are labeled with an asterisk. (Reproduced from Dhaka et al., 2006 with permission).

Chapter 3

STRUCTURAL-FUNCTIONAL CHARACTERISTICS OF TRP CHANNELS

Structural studies on TRP channels are poised for a quickened pace and rapid expansion. These structures enhance our understanding of TRP channel assembly and regulation. Continued technical advances in structural approaches promise a bright outlook for TRP channel structural biology (Li et al., 2011).

3.1. STRUCTURAL BIOLOGY OF TRP CHANNELS

Our understanding of the TRP complex membrane proteins made a huge leap forward through numerous high-resolution cryo-electron microscopy (cryo-EM) structures of near full-length TRP channels, beginning with TRP vanilloid 1 (TRPV1). Structures of near full-length ion channels are complemented nicely by structures from various soluble TRP channel domains (Goretzki et al., 2018).

TRP channel architecture is similar to other ion channels. Most members of the TRP channel superfamily share a low level of structural similarity, but some channels – such as TRPC3 and TRPC7, as well as TRPV5 and TRPV6 – are highly homologous to each other. Most of the

TRP channels are predicted of a six-transmembrane helix topology (S1 through S6) with a reentrant loop between S5 and S6 forming the channel pore that is a recurring structural motif. These channels tetramerize to a 24-helix functional protein complex (Figure 4).

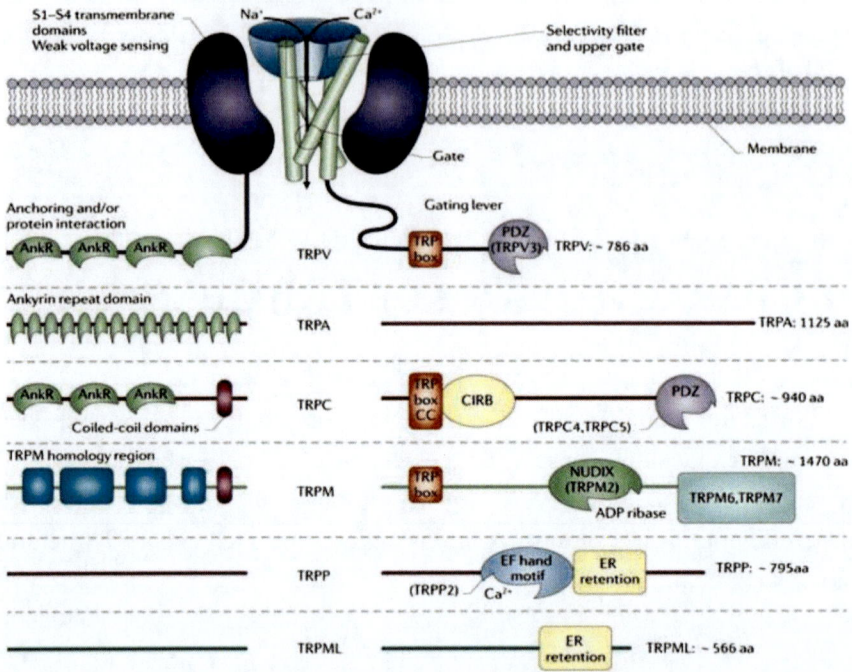

Figure 4. The six TRP cation families contain very different motifs in their amino and carboxyl termini. The TRPV, TRPA and TRPC families have amino terminal ankyrin repeat (AnkR) domains that are not present in other TRP channel subfamilies. The TRP box, which is found in the TRPV, TRPM and TRPC families, is thought to be involved in gating. TRP cation channel TRPP and TRPML proteins both have endoplasmic reticulum (ER) retention domains that may be due to their functional localization on intracellular organelles. aa, amino acids; CIRB, calmodulin/inositol-1,4,5-trisphosphate (Ins(1,4,5)P3) receptor binding domain; NUDIX, nucleoside diphosphate-linked moiety X; PDZ, acronym for postsynaptic density protein 95 (PSD95), Drosophila disc large tumor suppressor (DLGA) and zonula occludens protein 1 (ZO1). (Reproduced from Moran et al., 2011 with permission).

Many TRP channels form functional channels as homo-tetramers, although hetero-multi-merization is not uncommon. The latter phenomenon may have important implications in drug discovery as it is crucial for understanding the endogenous subunit composition of the ion channels so that TRP can be appropriately targeted with a pharmacological agent (Moran et al., 2011).

As observed for other ion channels, the TRP channel function is strongly influenced by large intracellular domains, and the responsiveness to functional modulators, e.g., regulation by phosphor-inositides, or inhibition by quaternary ammonium ions, and venom toxins, is conserved across ion channel families (Hellmich, Gaudet, 2014). Although all functional TRP channels are likely composed of four subunits (either homo- or hetero-tetramers), there is relatively low sequence homology between family members, and the overall structure of the channels can diverge significantly. Recently, the structures of several TRP channels have been elucidated with cryo-EM which is shown in Figure 5.

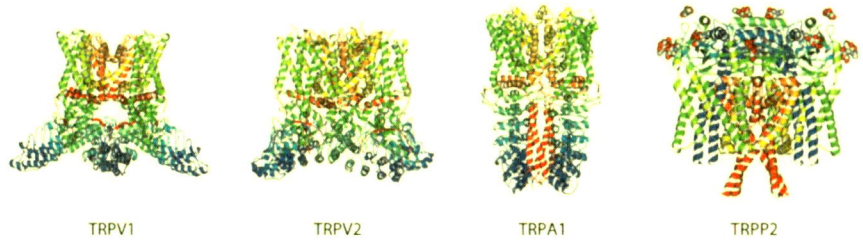

Figure 5. Comparison of recently elucidated TRP channel structures showing a side-on view of the channel indicated. (Reproduced from Moran, 2018 with permission).

Currently, structure models solved by cryo-EM are available for 48 TRP channels from 11 subfamilies (Figure 6) and the number of structures is still growing. Benefiting from the potential of cryo-EM to identify conformational heterogeneity by 2D and 3D classification methods, for almost all TRP channels, different conformational states were observed; e.g., open and closed conformations, and inhibited or activated states. Moreover, in many structures, the channels show association with lipids (Madej, Zeigler, 2018).

Comparisons between structures of different TRP channel subfamilies show similarities and, more importantly, differences that provided molecular insights into ion selectivity, gating, and activation. TRP channels share the tetrameric assembly, which is well preserved in detergent, amphipol, or reconstituted into nanodiscs. Due to the organization similarities of the membrane channel portion, in particular in the pore domain harboring the selectivity filter, comparisons of available structures gave profound insights into ion selectivity and permeation (Madej, Zeigler, 2018).

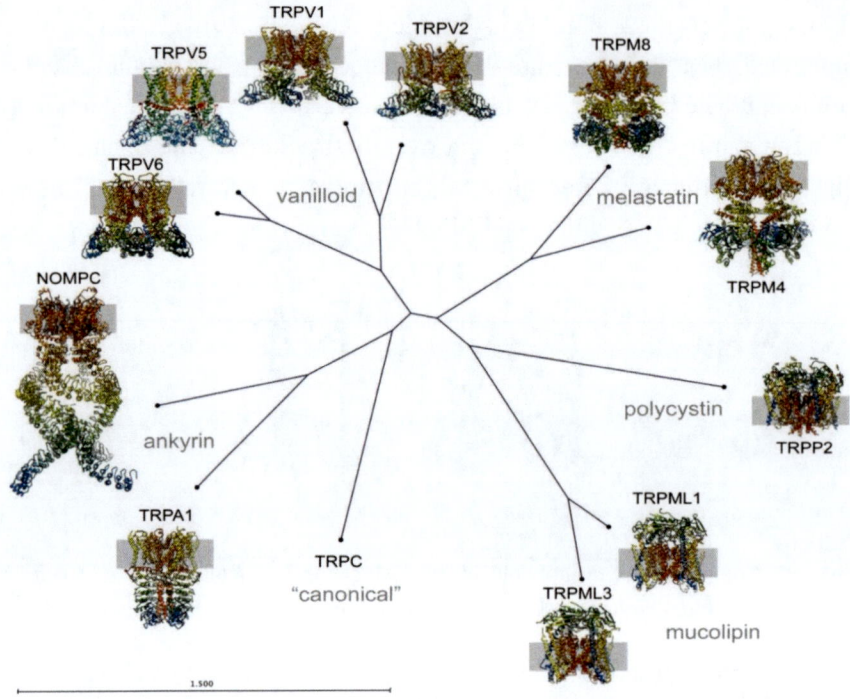

Figure 6. Phylogenetic similarity of TRP channels with resolved electron microscopy maps. The respective coordinates are shown as ribbon models, color ramping from N- (blue) to C-terminus (yellow). The gray rectangle indicates the extent of the membrane. (Reproduced from Madej, Zeigler, 2018 with permission).

It has been recently established in David Clapham laboratory (Duan et al., 2018a) that TRPM6 and TRPM7 are ubiquitously expressed with prominent roles in early embryonic development. Uniquely, these channels also include an active kinase domain. The functions of TRPM6 and TRPM7 are correlated with proteolytic cleavage of the kinase domain, which is then translocated to the nucleus to phosphorylate histones and regulate gene expression. TRPM7 contains both an ion channel and α-kinase. The channel domain comprises a non-selective cation channel with notable permeability to Mg^{2+} and Zn^{2+}. The structures reveal key residues for an ion binding site in the selectivity filter, with proposed partially hydrated Mg^{2+} ions occupying the center of the conduction pore. In high $[Mg^{2+}]$, a prominent external disulfide bond is found in the pore helix, which is essential for ion channel function. These results provide a structural framework for understanding the TRPM1/3/6/7 subfamily and extend the knowledge base upon which to study the diversity and evolution of TRP channels (Duan et al., 2018a).

Furthermore, they have found a cryo-EM structure of TRPC4 in its unliganded (apo) state to an overall resolution of 3.3 Å. The structure reveals a unique architecture with a long pore loop stabilized by a disulfide bond. Beyond the shared tetrameric six-transmembrane fold, the TRPC4 structure deviates from other TRP channels with a unique cytosolic domain. This unique cytosolic N-terminal domain forms extensive aromatic contacts with the TRP and the C-terminal domains (Duan et al., 2018c). TRPC4 has been characterized as a non-selective cation channel with moderate selectivity for Ca^{2+} over monovalent cations. Comparison of this TRPC structure with other TRP channel structures highlights some commonalities and differences.

In addition, this group of scientists has observed the electron cryo-microscopy structure of human TRPM4 in a closed, Na^+-bound, apo state at pH 7.5 to an overall resolution of 3.7 Å. Five partially hydrated sodium ions were proposed to occupy the center of the conduction pore and the entrance to the coiled-coil domain. They identified an upper gate in the selectivity filter and a lower gate at the entrance to the cytoplasmic coiled-coil domain. Intramolecular interactions exist between the TRP domain and the S4–S5 linker, N-terminal domain, and N and C termini. This TRPM4 channel is a widely distributed, calcium-activated, monovalent selective cation channel. Mutations in human TRPM4 result in progressive familial heart block (Duan et al., 2011 8b).

3.2. Physiological Functions of TRP Channels

The TRP channels have myriad functions in cell biology, including migration, proliferation and survival. They play roles in carcinogenesis, immunological processes, tissue homeostasis, and are prominently present in pathways of sensory transduction and electrolyte homeostasis. All TRP channels are cation permeable channels and exhibit at least some calcium conductance, exerting their function mostly by facilitating spatiotemporal calcium signaling events. In Particular, two TRP channels, TRPV5 and TRPV6, have evolved into highly selective ion channels that partake in the process of epithelial calcium transport, contributing significantly to the maintenance of human calcium homeostasis (van Goor et al., 2017).

The TRP channels play a plethora of physiological and pathological roles in response to various extracellular and intracellular stimuli, such as changes of temperature, pH, or osmolarity, injury, depletion of calcium stores, as well as volatile chemicals (odorants) and cytokines. Once activated, TRP channels, with homo- or hetero-tetrameric configurations, function as an integrator of several signaling pathways to elicit a serial of responses. Some TRPs are involved in sensory functions such as smell, taste, pain, and pheromone sensing. Some are responsive to temperature and may help to avoid tissue-damaging noxious temperatures. Some TRPs are sensitive to natural compounds or their ingredients that have been used in medical practice. It has been demonstrated that these TRP family members are involved in cell physiological functions and also in many hereditary and acquired diseases. Abnormalities in TRP channel function, as the result of alteration of protein expression levels, changes in channel properties, or changes in their myriad regulators, have been associated with numerous diseases ranging from chronic pain to cardiovascular disease, skeletal abnormalities, kidney diseases, brain disorders, and cancer, which provide numerous opportunities for therapeutic intervention (Li, 2017).

A large number of TRP channels can be activated directly by the binding of specific ligands, such as vanilloid compounds, endogenous lipids, ATP, and select medications, to binding pockets on the channel surface. Indirect activation that arises downstream of receptor-mediated signaling, either *via* G-protein coupled receptors (GPCR) or receptor tyrosine kinases (*trk*), also controls the activity of many TRP channels. Finally, several TRP channels are activated by biophysical mechanisms such as voltage, temperature and stretch. Interestingly, mechanical gating, caused by deformation of the plasma membrane, was proposed recently as a universal activation mechanism for all TRP channels (van Goor et al., 2017).

Chapter 4

A SHORT REVIEW OF TRP CHANNELS SUPERFAMILY

Extensive works on TRP channels for last decades, – especially on TRP gene knockout animals, obtained by deletion of individual *trp* genes in embryonic stem cells through homologous recombination, – have made it possible to identify TRP channel functions and their relationship to physiological and pathophysiological processes in the living organism. TRPs are probably expressed in all cells of human body. It was therefore not unexpected that TRP channels are involved in several, still not well-understood diseases and have therefore triggered a huge hope for the development of new drug targeting these channels (Flockerzi, Nilius, 2014). Significant interest in the therapeutic utility of TRP channels persists across disease areas despite these challenges because of the compelling biology. Therefore, not surprisingly, there has been significant pharmaceutical interest in targeting these fascinating channels (Moran, 2018).

4.1. TRPA Subfamily

The transient receptor potential 1 (TRPA1) is the only member of the ankyrin subfamily so far identified in mammals and constitutes its own subfamily. The TRPA1 protein, originally called ANKTM1 (ankyrin-like with transmembrane domains protein 1), was first identified in human fetal lung fibroblasts as a transformation-associated gene product. Later studies showed that TRPA1 was expressed in sensory neurons of dorsal root ganglion (DRG), nodose ganglion (NG) and trigeminal ganglion (TG) neurons and therefore, was crucial for the comprehension of the sensory functions, highlighting it as a putative component for the propagation of noxious and inflammatory stimuli (de Andrade et al., 2012).

TRPA1 is a nonselective cation channel with six putative transmembrane segments (S1–S6), intracellular N- and C-termini, and a pore loop between S5 and S6. The N-terminus contains between 14 and 18 ankyrin repeats that probably are important for protein–protein interactions and insertion of the channel into the plasma membrane. Because of the unusually large N-terminal Ankyrin repeat domain, it is also possible that TRPA1 is involved in mechanosensation, in which the N-terminal could act as a link between mechanical stimuli and channel gating. The N-terminal region contains a large number of cysteines, some of which can form a complex network of protein disulfide bridges within and between monomers (Figure 7). The N- and C-termini have been suggested to contain binding sites for Ca^{2+} that can both sensitize and desensitize TRPA1 (Zygmunt, Högestätt, 2014).

Nowadays, it is well known that TRPA1 is a sensor for chemical irritants (allicin, carvacrol, cinnamaldehyde, gingerol, mustard oil, thymol, wasabi) and a major contributor to chemo-nociception. Key residues involved in irritant detection are solvent accessible and lie within a putative allosteric nexus converging on an unpredicted TRP-like domain, suggesting a structural basis in which TRPA1 functions as a sensitive, low-threshold electrophilic receptor (Paulsen et al., 2015).

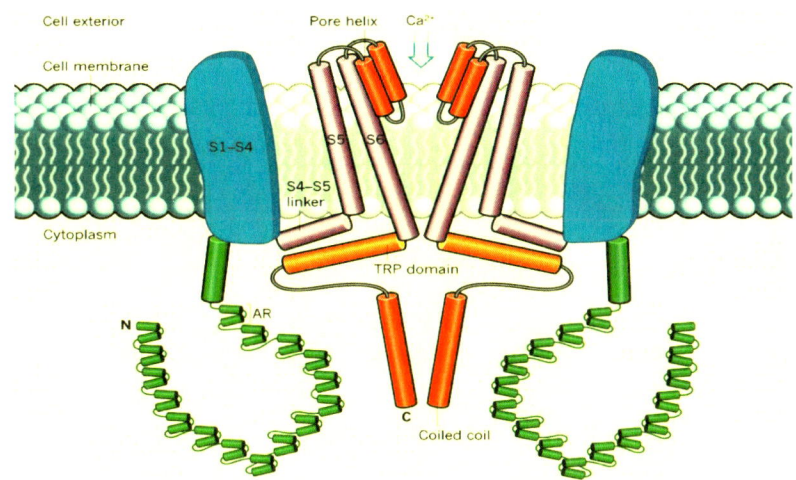

Figure 7. A schematic of key features of the ion channel TRPA1, which mediates the cellular influx of cations such as calcium ions (Ca^{2+}). Two TRPA1 subunits are shown, although the channel is comprised of four. Each subunit contains six membrane spanning α-helical domains, S1–S6. Two pore helices link S5 and S6 at the extracellular surface, where a constriction regulates influx of Ca^{2+}. The helical TRP domain is part of a second, lower constriction. Sixteen ankyrin repeats (ARs) at the amino-terminal end (N) of the subunit cover a carboxy-terminal (C) coiled-coil structure, providing a large cytoplasmic surface for interactions with noxious agents. It is probable that the molecular interactions of ligands with ARs lead to conformational changes, conveyed through the S4–S5 linker structure, that open the channel. (Reproduced from Clapham, 2015 with permission).

TRPA1 is a promiscuous chemical nocisensor that is also involved in noxious cold and mechanical sensation. It is present in a subpopulation of Aδ- and C-fiber nociceptive sensory neurons as well as in other sensory cells including epithelial cells. In primary sensory neurons, Ca^{2+} and Na^+ flowing through TRPA1 into the cell cause membrane depolarization, action potential discharge, and neurotransmitter release both at peripheral and central neural projections. In addition to being activated by cysteine and lysine reactive electrophiles and oxidants, TRPA1 is indirectly activated by pro-inflammatory agents *via* the phospholipase C signaling pathway, in which cytosolic Ca^{2+} is an important regulator of channel gating (Zygmunt, Högestätt, 2014).

4.2. TRPC Subfamily

The TRP-canonical (TRPC) subfamily, which consists of seven members (TRPC1–TRPC7), are Ca^{2+}-permeable cation channels that are activated in response to receptor-mediated phosphatidyl-inositol biphosphate (PIP2) hydrolysis *via* store-dependent and store independent mechanisms (Ong et al., 2014). This TRPC subfamily discovered almost 25 years ago (Wes et al., 1995; Zhu et al., 1995), is unique in that its members are not only responsible for agonist-activated nonselective cation currents, but in that they also participate in the so-called slow sustained mode of Ca^{2+} signaling, which requires sustained elevations of intracellular Ca^{2+}. This subfamily differs from the other TRP subfamilies in that it fulfills two different types of function: (1) membrane depolarization, resembling sensory transduction TRPs, and (2) mediation of sustained increases in intracellular Ca^{2+} (Birnbaumer, 2009).

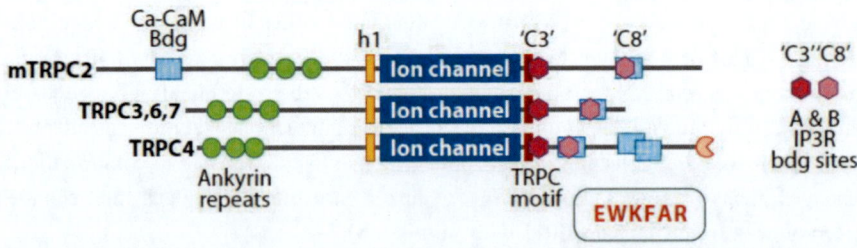

Figure 8. Structural domains of TRPC channels. All seven members of this subfamily are structurally similar having six transmembrane helices, a putative hydrophobic pore forming loop, three to four Ankyrin repeats, coiled-coil domains in the N- and C-terminus, a C-terminal proline rich region, a Calmodulin/IP3 binding region and the TRP motif. They have the so-called TRP domain immediately after the ion channel domain, which includes three variations of the Glu-Trp-Lys-Phe-Ala-Arg (EWKFAR) TRP box. (Adapted from Birnbaumer, 2009).

The seven members of this subfamily are commonly divided into four groups based on sequence homology (I) TRPC1; (II) TRPC2; (III) TRPC3, TRPC6, TRPC7; and (IV) TRPC4 and TRPC5. TRPC channels have been shown to form both hetero-tetramers and homo-tetramers within the TRP channel superfamily, with different members having certain preferences, for example, TRPC1 forms physiologically relevant functional channels with several TRPC channels including TRPC4, as well as, TRPV1, and TRPP2 (Samanta et al., 2018).

The advances in cryo-EM enable the high resolution structure determination of TRP ion channels, including the TRPC subfamily. These structures shed light on the architecture of this receptor-activated TRPC channel and provide a structural basis in understanding how the ion conduction pathway is built and the gating mechanism (Figure 8). The findings have broad implications for understanding the structural basis underlying calcium homeostasis, epitope antibody selectivity, and the diverse functional and physiological roles of this ion channel family. Although TRPCs has wide pharmaceutical applications in treatment of various diseases, drug development specifically targeting TRPCs has been limited due to the lack of understanding of its molecular activation mechanisms and the relationships between structural properties and the physiological conditions. These structures provide the starting point for future investigation on these important molecules (Li et al., 2019).

A growing body of evidence suggests a prime contribution of TRPC channels in regulating fundamental neuronal functions, including second messengers and store depletion. TRPC channels have been recently shown to be associated with neuronal development, proliferation and differentiation. In addition, TRPC channels have also been suggested to have a potential role in regulating neurosecretion, long-term potentiation, and synaptic plasticity. During the past years, numerous seminal discoveries relating TRPC channels to neurons have constantly emphasized on the significant contribution of this group of ion channels in regulating neuronal functions (Bollimuntha et al., 2009). The Ca^{2+} influx mediated by TRPC channels generates distinct intracellular Ca^{2+} signals that regulate downstream signaling events and consequent cell functions. This requires localization of TRPC channels in specific plasma membrane microdomains and precise regulation of channels function which is coordinated by various scaffolding, trafficking, and regulator proteins (Ong et al., 2014).

Finally, TRPC channels have emerged as central players in various physio-pathological processes in non-neuronal systems. In particular, in the kidney, cardiovascular system, and lung, the three major organs in which the functions of TRPC channels have been most extensively studied in the past years. Mutations in these proteins are frequently associated with human diseases. As more information from the *in vivo* role of TRPC channels in animal models and clinical data from patients carrying mutations become available, our knowledge of the role of TRPC channels in disease pathogenesis will expand considerably (Tai et al., 2017).

4.2.1. TRPC1

As stated above, the TRPC1 ion channel was one of the first mammalian TRP channels to be cloned in *Drosophila* (Zhu et al., 1995). In humans, it is encoded by the TRPC1 gene located in chromosome 3. The protein is predicted to consist of six transmembrane segments with the N- and C-termini located in the cytoplasm. The extracellular loop connecting transmembrane segments 5 and 6 participates in the formation of the ionic pore region. Inside the cell, TRPC1 is present in the endoplasmic reticulum, plasma membrane, intracellular vesicles, and primary cilium, an antenna-like sensory organelle functioning as a signaling platform. In human and rodent tissues, it shows an almost ubiquitous expression. TRPC1 interacts with a diverse group of proteins including ion channel subunits, receptors, and cytosolic proteins to mediate its effect on Ca^{2+} signaling. It primarily functions as a cation nonselective channel within pathways controlling Ca^{2+} entry in response to cell surface receptor activation. Through these pathways, it affects basic cell functions, such as proliferation and survival, differentiation, secretion, and cell migration, as well as cell type-specific functions such as chemotropic turning of neuronal growth cones and myoblast fusion. The biological role of TRPC1 has been studied in genetically engineered mice where the *Trpc1* gene has been experimentally ablated. Although these mice live to adulthood, they show defects in several organs and tissues, such as the cardiovascular, central nervous, skeletal and muscular, and immune systems. Genetic and functional studies have implicated TRPC1 in diabetic nephropathy, Parkinson's disease, Huntington's disease, Duchenne muscular dystrophy, cancer, seizures, and Darier–White skin disease (Dietrich, 2014; Nesin, Tsiokas, 2014).

Together, in the near future agonists and antagonists of TRPC1 and TRPC4/5 channels as well will provide valuable hints to comprehend the functional importance of these ion channels in native cells and *in vivo* animal models. Importantly, human diseases and disorders mediated by these proteins can be studied using these compounds to perhaps initiate drug discovery efforts to develop novel therapeutic agents (Rubaiy, 2019).

4.2.2. TRPC2

The TRPC2 was the second ortholog of the *Drosophila* TRP gene to be identified. Whereas full-length TRPC2 transcripts have been cloned in a number of species including mice, rats and New World monkeys, and in a number of other vertebrates, TRPC2 is lightly expressed in the rodent vomero-nasal organ (VNO). It is also detectable at the protein level in murine erythroblasts, sperm, and brain and has been detected in other tissues by the reverse transcription polymerase chain reaction (RT-PCR). Its activation by diacylglycerol (DAG) and by erythropoietins has been described in greatest detail, and inhibition by Ca^{2+}-calmodulin has been reported. The major demonstrated functions of TRPC2 are regulation of pheromone-evoked signaling in the rodent VNO, regulation of erythropoietin-stimulated calcium influx in murine erithroid cell, and the zona pellucida sperm-binding protein 3 (ZP3)-evoked calcium influx into sperm. Depletion of TRPC2 in knockout mice resulted in clones in behavior including altered sex discrimination and lack of male-male aggression. The red blood cells of TRPC2 knockout mice showed increased mean corpuscular volume, mean corpuscular hemoglobin, hematocrite, and reduced mean corpuscular hemoglobin concentration. TRPC2-depleted red cells were resistant to oxidative stress-induced hemolysis (Miller, 2014). While the human *Trpc2* is a pseudogene and does not form a functional channel (Wes et al., 1995; Zhu et al., 1995), TRPC2 in other mammals such as rat, bovine, and mouse, forms functional channels in different cell types and tissues, such as the VNO, testis, spleen, and liver. TRPC2 has been reported to co-localize with anoctamin 1 in the vomero-nasal epithelium, although the interaction between the two proteins has not been confirmed using other techniques such as immune-precipitation, fluorescence resonance energy transfer (FRET) and total internal reflection fluorescence (TIRF) microscopy (Ong et al., 2014).

4.2.3. TRPC3

TRPC3 represents one of the first identified mammalian relative of the *Drosophila* TRP gene product. Despite extensive biochemical and biophysical characterization as well as ambitious attempts to uncover its physiological role in native cell systems, the channel protein still represent a rather enigmatic member of the TRP superfamily. TRPC3 is significantly expressed in the brain and heart and appears of (patho) physiological importance in both non-excitable cells and excitable cells, being potentially involved in a wide spectrum of Ca^+ signaling mechanisms. TRPC3 cation channels display unique getting and regulatory properties that allow for recognition and integration of multiple input stimuli including lipid mediators, cellular Ca^{2+} gradients, as well as redox signals. Physiological/ pathophysiological functions of the highly versatile cations channel protein are as yet incompletely understood. Its ability to associate in a dynamic manner with a variety of partner proteins enables TRPC3 to serve coordination of multiple downstream signaling pathways and control of divergent cellular functions. A crucial role of this signaling molecule in cardiovascular and neuronal pathologies as well as its significance as a therapeutic target is likely (Lichtenegger, Groschner, 2014).

TRPC3 channels are expressed in the endothelium and/or smooth muscle of specific intact arteries, such as mesenteric, cerebral and myometrial, where they are critical for the control of vascular tone, and show altered activity in development and disease. In artery endothelium, TRPC3 contributes to endothelium-derived hyperpolarization and nitric oxide-mediated vasodilatation. In artery smooth muscle, TRPC3 contributes to constrictor mechanisms. In both endothelium and smooth muscle, TRPC3 contributes to function *via* caveolae-caveolin dependent and independent mechanisms. In different cell types and states, like other TRP channels, TRPC3 can form complexes with other TRP proteins and associated channels and accessory proteins, including those associated with endo-(sarco)-plasmic reticulum (ER/SR), thereby facilitating Ca^{2+} channel activation and/or refilling ER/SR Ca^{2+} stores. The diversity of TRPC3 interactions with other vascular signaling components is a potential target for artery specific control mechanisms (Grayson et al., 2017).

4.2.4. TRPC4

TRPC4 proteins comprise six transmembrane domains a putative pore-forming region, and an intra-cellular located amino- and carboxy-terminus. Among eleven splice variants identified so far, TRPC4$_\alpha$ and TRCP4$_\beta$ are most abundantly expressed and functionally characterized. TRPC4 is expressed in various organs and cell types including the soma and dendrites of numerous types of neurons; the cardiovascular system including endothelial, smooth muscle, and cardiac cells; myometrial and skeletal muscle cells; kidney; and immune cells such as mast cells. Both recombinant and native TRPC4-contaning channels differ tremendously in their permeability and other biophysical properties, pharmacological modulation, and mode of activation depending on the cellular environment. They vary from inwardly rectifying store-operated channels, with a high Ca^{2+} selectively to non-store-operated channels predominantly carrying Na^+ and activated by $G_{\alpha q}$ - and/or $G_{\alpha i}$ receptors with a complex U-shaped current voltage relationship. Thus, individual TRPC4-contacting channels contribute to agonist-induced Ca^{2+}entry directly or indirectly *via* depolarization and activation of voltage-gated Ca^{2+}-channels. The biological relevance of TRPC4 contacting channels was demonstrated by knockdown of TRPC4 expression numerous native system including gene expression, cell differentiation and proliferation, formation of myotubes, and axonal regeneration. Studies of TRPC4 single and TRPC compound knockout mice uncovered their role for the regulation of vascular tone, endothelial permeability, gastrointestinal contractility and motility, neurotransmitter release, and social exploratory behavior as well as for excite-toxicity and epileptogenesis. Recently a single-nucleotide polymorphism (SNP) in the *Trpc4* gene was associated with a reduced risk for experience of myocardial infarction (Freichel et al., 2014).

TRPC4 is most closely related to TRPC5, sharing 65% amino acid identity, but both proteins diverge in the last 220 amino acids. There is general consensus that TRPC4 forms a store-operated channel (SOC) even though it has been shown to form constitutively active or store-independent channels in some studies (Ong et al., 2014; Venkatachalam, Montell, 2007; Venkatachalam et al., 2014).

4.2.5. TRPC5

Human canonical transient receptor channel 5 (TRPC5) has been cloned from the Xq23 region on chromosome X as a suspect in non-syndromic mental retardation. TRPC5 is a Ca^{2+} permeable cation channel predominantly expressed in the CNS, including the hippocampus, cerebellum, amygdala, sensory neurons, and retina. It also shows more restricted expression in the periphery, notably in activated by reception coupled to G_q and phospholipase C and/or G_i proteins, but TRPC5 may also gate in a store-depended manner, which requires other partner proteins such TRPC1, stromal interaction molecule 1 (STIM1), and Orai1. There is an impressive array of other activation of TRPC5 channels, such as nitric oxide, lysophospholipids, sphingosine-1-phospate, reduced thioredoxin. Moreover, TRPC5 shows constitutive activity, and it is responsive to membrane stretch and cold. Thus, TRPC5 channels have significant potential for synergistic activation and may serve as an important focal point in Ca^{2+} signaling and electrogenesis. Moreover, TRPC5 functions in partnership with about 60 proteins, including TRPC1, TRPC4, calmodulin, IP_3 receptors, NHERF, neuronal calcium sensor-1 (NCS-1), junctate, stathmin 2, Ca^{2+} binding protein 1, caveolin and SESTD1, while its desensitization is mediated by both protein kinases A and C. TRPC5 has a distinct voltage dependence shared only wide its closest relative TRPC4. Its unique N-shaped activation curve underlined by intracellular Mg^{2+} block seems to be perfectly 'shaped' to trigger action potential discharge, but not to grossly interfere with the action potential shape. The range of biological functions of TRPC5 channels is also impressive, from neurotransmission to control of axon guidance and vascular smooth muscle cell migration and contractility. Recent studies of *Trpc5* gene knockout begin to uncover its roles in fear, anxiety, seizures, and cold sensing (Ong et al., 2014; Rubaiy, 2019; Zholos, 2014).

The precise activation mechanisms of TRPC4 and TRPC5 channels have remained largely elusive for a long time, as they were considered as being insensitive to the second messenger DAG in contrast to the other TRPC channels. Recent findings, however, indicate that the C-terminal interactions with the scaffolding proteins Na^+/H^+ exchanger regulatory factor 1 and 2 (NHERF1 and NHERF2) dynamically regulate the DAG sensitivity of the TRPC4 and TRPC5 channels. Interestingly, the C-terminal NHERF binding suppresses, while the dissociation of NHERF enables, the DAG sensitivity of the TRPC4 and TRPC5 channels. This leads to the assumption that all of the TRPC channels are DAG sensitive. The identification of the regulatory function of the NHERF proteins in the TRPC4/5 – NHERF protein complex offers a new starting point to get deeper insights into the molecular basis of TRPC channel activation (Mederos y Schnitzler et al., 2018).

Interesting data were found in the Pertovaara's lab. They studied whether TRPC4/C5 channels in the central nucleus of amygdala (CeA) in rodent's brain that are known to have anxiogenic properties contribute to the maintenance of sensory or affective aspects of pain in an experimental model of peripheral neuropathy. Rats with a spared nerve injury (SNI) model of neuropathy in the left hind limb had a chronic cannula for microinjections of drugs into the right amygdala or the internal capsule (a control site). Sensory pain was assessed by determining mechanical hypersensitivity with calibrated monofilaments and affective pain by determining aversive place-conditioning. Amygdaloid treatment with ML-204, a TRPC4/C5 antagonist, produced a dose-related (5–10 μg) antihypersensitivity effect, without obvious side-effects. Additionally, amygdaloid administration of ML-204 reduced affective-like pain behavior. In the internal capsule, ML-204 had no effect on hypersensitivity or affective-like pain in SNI animals. In healthy controls, amygdaloid administration of ML-204 failed to influence pain behavior induced by mechanical stimulation or noxious heat. The results indicate that the amygdaloid TRPC4/C5 channels contribute to maintenance of pain hypersensitivity and pain affect in neuropathy. It is hypothesized, thus, that the amygdaloid TRPC4/C5 might exert a role in comorbidity between affect and chronic pain and thereby, they might provide therapeutic targets in affective disorder associated chronic neuropathic pain conditions (Wei et al., 2015).

4.2.6. TRPC6

Canonical transient receptor potential 6 (TRPC6) proteins assemble into hetero-multimeric structures forming non-selective cation channels. In addition, many TRPC6-interacting proteins have been identified like some enzymes, channels, pumps, cytoskeleton-associated proteins, immunephilins, or cholesterol binding proteins, indicating that TRPC6 are engaged into macromolecular complexes. Depending on the cell type and the experimental conditions used, TRPC6 activity has been reported to be controlled by diverse modalities. For instance, the second messenger DAG, store-depletion, the plant extract hyperforin or H_2O_2 have all been shown to trigger the opening of TRPC6 channels. A well-characterized consequence of TRPC6 activation is the elevation of the cytosolic concentration of Ca^{2+}. This latter response can reflect the entry of Ca^{2+} through open TRPC6 channels but it can also be due to the Na^+/Ca^{2+} exchanger (operating in its reverse mode) or voltage-gated Ca^{2+} channels (recruited in response to a TRPC6-mediated depolarization). Although TRPC6 controls a diverse array of biological functions in many tissues and cell types, its pathophysiological functions are far from being fully understood (Bouron et al., 2016).

TRPC6 is likely to play a number of physiological roles which are confirmed by the analysis of a *Trpc6* $^{-/-}$ mouse model. In smooth muscle Na^+ ions influx through TRPC6 channels, and activation of voltage-gated Ca^{2+} channels by membrane depolarization is the driving force for contraction. Permeability of pulmonary endothelial cells depends on TRPC6 and induces ischemia-reperfusion edema formation in the lungs. TRPC6 was also identified as an essential component of the slit diaphragm architecture of kidney podocytes and play an important role in the protection of neurons after cerebral ischemia. Other function especially in immune and blood cells remain elusive. Recently identified TRPC6 blockers may be helpful for therapeutic approaches in diseases with highly activated TRPC6 channel activity (Dietrich, Gudermann, 2014)

Recent evidence have clearly shown an involvement of some TRP channels (TRPA1, TRPC4, TRPC6, TRPV1, TRPV4, TRPM2 and TRPM8) expressed in the lung and respiratory airways for toxin sensing and as mediators of lung inflammation and associated diseases like asthma, chronic obstructive pulmonary disease (COPD), lung fibrosis and edema formation (De Logu et al., 2016; Dietrich et al., 2017). Specific modulators of these channels may offer new therapeutic options in the future and will endorse strategies for a causal, specifically tailored treatment based on the mechanistic understanding of molecular events induced by lung-toxic agents (Dietrich, 2019).

4.2.7. TRPC7

The TRPC7 channel is the seventh member of the mammalian TRPC channel family. TRPC7 mRNA protein, and channel activity have been detected many tissues and organs from the mouse, rats and human. TRPC7 has high sequence homology with TRPC3 and TRPC6 and all three channels are activated by membrane receptors that couple to isoforms of phospholipase C (PLC) and mediate non-selective cation currents. TRPC7, along with TRPC3 and TRPC6, can be activated by direct exogenous application of DAG analogues and by pharmacological maneuvers that increase endogenous DAG in cells. TRPC7 shows distinct properties of activation, such as constitutive activity and susceptibility to negative regulation by extracellular Ca^{2+} and by protein kinase C. TRPC7 can form hetero-multimers with TRPC3 and TRPC6. Although TRPC7 remains one of the least studied TRPC channels, its role in various cell types and physiological and pathophysiological conditions is beginning to emerge (Zhang, Trebak, 2014).

Altered TRPC7 channel expression and activity are observed in various diseases including malignant breast cancer tumors. Regulation of multiple TRPC genes has been shown to be mediated by the Rho-associated protein kinase pathway. Rho-kinase activity has been associated with metastasis of esophageal cancer cell lines and inhibition of this pathway using the pharmacological inhibitor Y-27632 prevented the growth and invasiveness of these cancer cells. Interestingly the Rho-kinase inhibitors attenuated expression of TRPC1 and another TRP family member TRPV2. This study suggests a potential protective role of TRPC7 in cancer cell growth and progression TRPC7 expression has also been shown to be up-regulated in animal models of disease. In Dahl salt sensitive rats, TRPC7 mRNA was increased in the failing myocardium of these animals. This increased expression correlated with an increased level of apoptosis as assessed by TUNEL staining. Heart failure in these mice was suggested to be contributed by angiotensin II-induced Ca^{2+} entry through activation of TRPC7 and subsequent myocardial apoptosis. In a mouse model of pilocarpine induced *status epilepticus*, TRPC7 was shown to mediate the initiation of acute seizures, as observed by the reduction in pilocarpine induced gamma wave activity in TRPC7 knockout animals. The mechanism by which TRPC7 mediates seizure induction was characterized in brain slices derived from the hippocampal region *cornu ammonis* (CA3), a central region involved in seizure generation. TRPC7 was shown to generate spontaneous epileptiform burst firing, these signals of seizure induction were initiated at the synaptic level where TRPC7 was shown to be involved in the potentiation of these signals in the CA3 synapses and Schaffer collateral-CA1 synapses. This study using TRPC7 knockout animals was one of the first studies to provide insight into the *in vivo* function of this enigmatic channel protein (Zhang et al., 2016).

4.3. TRPM Subfamily

The members of the TRPM (melastatin) subfamily are divided into 4 groups on the basis of sequence homology: TRPM1/3, TRPM2/8, TRPM4/5, and TRPM 6/7. Ca^{2+} permeability in the TRPM family ranges from impermeable to Ca^{2+} (TRPM4 and 5) to highly Ca^{2+} permeable (TRPM3, 6 and 7). Unlike the other TRP channels, TRPM channels lack the N-terminal ankyrin repeats (Gees et al., 2012).

Considering the level of sequence and structural similarity between members of TRPM subfamily, it is not surprising that there is also cross-talk by open and close states within the group. A combination of traditional mutagenesis methods and some of the more recent developments will provide further insights, many of which will be widely applicable to the whole TRP channels. There are now 4 unique TRPM structures from over 20 protein data bank (PDB) entries (Figure 9), which have presented many unprecedented findings in regard to the diversity of the structure, ion permeation and gating mechanism (Chen et al., 2019).

The TRP box is situated in the C-terminal domain of the TRPM family, though there are no ankyrin repeats in the N-terminus (Figure 10). According to the latest data obtained using 3D cryo-structure electronic microscopy, the TRP domain is not a continuous α-helix, but rather, has a break at about two-thirds of the helical structure length, thus, dividing the TRP domain into two segments (Winkler et al., 2017). The N-terminal part of all TRPM proteins is usually longer, by approximately 300–400 amino acids (aa), than the same regions in TRPC and TRPV families. In this region, there is also a huge, approximately 700 aa TRPM-homology domain (Figure 10). The C-terminal sequences of the TRPM family display high variabilities, differing from as much as 1000 to 2000 aa (Hantute-Ghesquier et al., 2018).

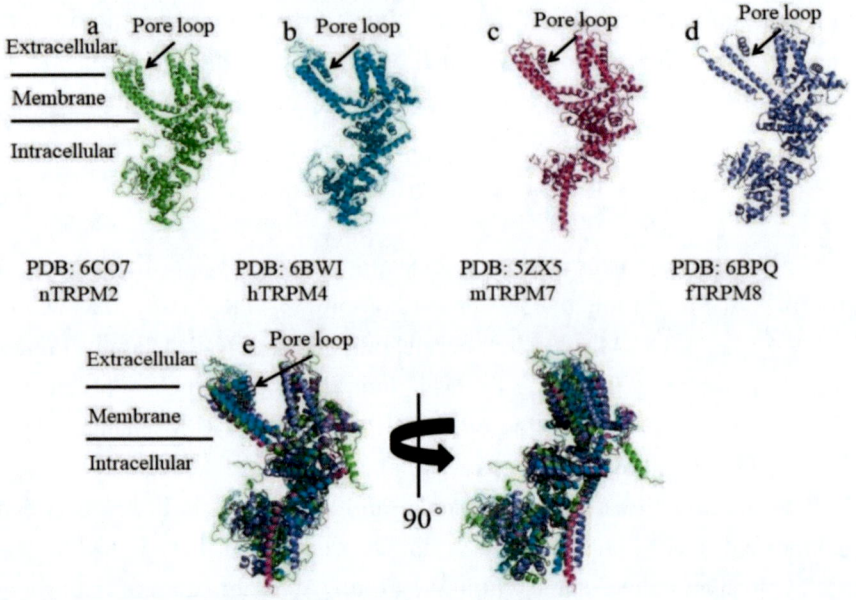

Figure 9. Structures of TRPM channels. (A) Cartoon representation of a *n*TRPM2 (*n: Nematostella vectensis*) protomer. (B) Cartoon representation of a *h*TRPM4 (*h: Homo sapiens*) protomer. (C) Cartoon representation of a *m*TRPM7 (*m: Mus musculus*) protomer. (D) Cartoon representation of a *f*TRPM8 (*f: Ficedula albicollis*) protomer. (E) Comparison of *n*TRPM2 (green), *h*TRPM4 (cyan), *m*TRPM7 (magentas) and *f*TRPM8 (slate). (Reproduced from Chen et al., 2019 with permission).

TRPM channels express in wide range of organ systems throughout our body, such as immune and cardiovascular systems, which are essential guarantee for human health. Corresponding to their broad expression, TRPM channels play a significant role in cell development and relevant diseases which are urgent to be solved. For instance, TRPM4 and TRPM5 have been found as a modulator in taste cells, TRPM7 is involved in mammalian homeostasis, and TRPM8 is important in cold sensation (Chen et al., 2019).

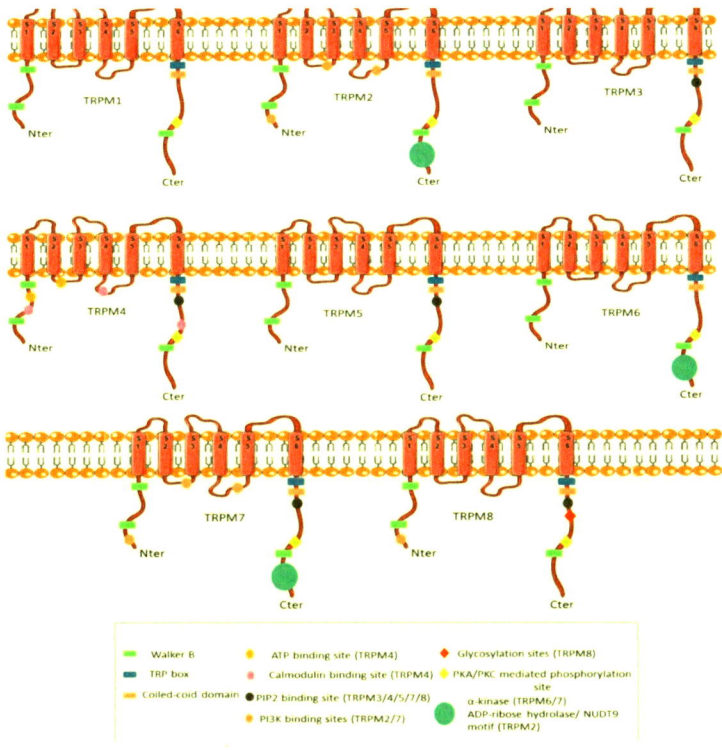

Figure 10. General structure of the TRPM family of channels (reproduced from Hantute-Ghesquier et al., 2018).

4.3.1. TRPM1

It has recently revealed that TRPM1, the founding member of the melanoma-related TRPM subfamily, is required for photoreceptors in mouse retinal ON-bipolar cells. Further experiments have demonstrated that TRPM1 is component of the transduction cation channel negatively regulated by the metabotropic glutamate receptor 6 (mGluR6) cascades in ON-bipolar cells through a reconstitution experiment using Chinese hamster ovary (CHO) cells expressing TRPM1, mGluR6 and Goα. Furthermore, human TRPM1 mutations are associated with congenital stationary night blindness (CSNB) patients whose lack rod functions and suffer from night blindness starting in early childhood. In addition to the function of transduction cation channel, TRPM1 is one of the retinal autoantigens in some paraneoplastic retinopathy (PR) associated with retinal ON-bipolar cell dysfunction (Irie, Furukawa, 2014).

The available data clearly indicate that TRPM1 proteins in the dendritic tips of ON-bipolar cells are embedded in a large multi-protein complex, akin to the archetypical TRP channels of *Drosophila* photoreceptors that also interact with a large number of different proteins in a multi-protein complex that has been termed signalplex or transducisome (Schneider et al., 2015).

4.3.2. TRPM2

TRPM2 is the second member of the TRPM subfamily of cation channels. The protein is widely expressed including in the brain, immune system, endocrine cells, and endothelia. It embodies both ion channel functionality and enzymatic ADP-ribose (ADPr) hydrolase activity. TRPM2 is a Ca^{2+}-permeable non-selective cation channel embedded in the plasma membrane and/or lysosomal compartments that in primarily activated in a synergistic fashion by intracellular ADP-ribose and Ca^{2+}. It is also activated by reactive oxygen and nitrogen species (ROS/NOS) and enhanced by additional factors, such as cyclic ADPr and NAADP, while inhibited by permeating protons (acidic pH) and adenosine monophosphate (AMP). Activation of TRPM2 leads to increase in intracellular Ca^{2+} levels, which can serve signaling roles in inflammatory and secretory cells through release of vesicular mediators (e.g., cytokines, neurotransmitters, insulin) and in extreme cases can include apoptotic and necrotic cell death under oxidative stress (Faouzi, Penner, 2014).

TRPM2 has recently emerged as a likely candidate for the detector of non-noxious warmth, as it is expressed in sensory neurons and mice show deficits in the detection of warmth when TRPM2 is genetically deleted. It plays a role in regulating body temperature, in the immune system and cancer, in pain and insulin secretion (Tan, McNaughton, 2018).

TRPM2 is also highly expressed in many types of cancer, including breast, prostate, and pancreatic cancer, melanoma, leukemia, and neuroblastoma, suggesting TRPM2 promotes tumor survival. When TRPM2-mediated calcium influx is inhibited, mitochondria are dysfunctional, cellular bioenergetics are reduced, production of the reactive oxygen species (ROS) is increased, and autophagy and DNA repair are impaired, decreasing tumor growth and increasing chemotherapy sensitivity. Inhibition of TRPM2 expression or function results in decreased tumor proliferation and/or viability in many malignancies including breast, gastric, pancreatic, prostate, head and neck cancers, melanoma, neuroblastoma, and T-cell and acute myelogenous leukemia. However, in a small number of malignancies, activation of TRPM2 rather than inhibition has been reported to reduce tumor cell survival. Mechanisms through which TRPM2 mediates these effects are under investigation and TRPM2 is considered as an exciting potential target in therapy of many cancers as further evidence of its important function in cell survival is forthcoming and controversies in its essential roles are clarified (Miller, 2019).

4.3.3. TRPM3

Like most other member of the TRP superfamily, the *Trpm3* gene encodes proteins that form cation-permeable ion channels on the plasma membrane. However, TRPM3 proteins have several unique features that set them apart from the member of this diverse family. The *Trpm3* gene encodes for surprisingly large number of isoforms generated mainly by alternative splicing. Only for two of the eight sides at which sequence diversity is generated, the functional consequence has been elucidated, one is leading to non-functional channels, and the other one is profoundly affecting the ionic selectivity. Over the past years, important progress has been made in discovering pharmacological tools to manipulate TRPM3 channel activity. These substances have facilitated the identification of endogenously expressed functional TRPM3 in nociceptive neurons, pancreatic beta cells, and vascular smooth muscle cells, among others. TRPM3 channels, which themselves are temperature sensitive, thus have been implicated in sensing noxious heat, in modulating insulin release and in secretion of inflammatory cytokines, however in many issues where TRPM3 proteins are known to be expressed, no functional role has been identified for these channels so far. Because of the availability of adequate pharmacological and genetic tools, it is expressed that future investigation on TRPM3 channels will unravel important new aspects and functions of these channels (Oberwinkler, Philipp, 2014).

Recent data show that TRPM3 channels are expressed in adipocytes, pancreatic beta-cells and in the kidney, eye, and brain (in somatosensory neurons) as well as in the pituitary gland. TRPM3 channel stimulation is likely to play a particular, tissue-specific role in these tissues and cell types and some new functional research studies have begun to unravel the physiological roles of TRPM3 channels (Thiel et al., 2017).

Just recently it has been shown that TRPM3 expression in mesenteric arteries is restricted to perivascular nerves. TRPM3-dependent vasodilatation was partially inhibited by a cocktail of K^+ channel blockers, and not mediated by β-adrenergic signaling. Contrary to what was found in aorta, the neurosteroid pregnenolone sulfate (PS) dilates mesenteric arteries, partly *via* an activation of TRPM3 that triggers the *calcitonin gene-related peptide* (CGRP) release from perivascular nerve endings and a subsequent activation of K^+ channels in vascular smooth muscle cells (VSMC). The authors propose that TRPM3 is implicated in the regulation of the tone of resistance arteries and that its activation by yet unidentified endogenous damage-associated molecules lead to protective vasodilation responses in mesenteric arteries (Alonso-Carbajo et al., 2019).

4.3.4. TRPM4

TRPM4 is a calcium-activated, phosphatidylinositol-4,5-bisphosphate (PtdIns(4,5)P2)–modulated, non-selective cation channel. In terms of structure, TRPM4 consists of multiple transmembrane and cytosolic domains, which assemble into a three-tiered architecture. The N-terminal nucleotide-binding domain and the C-terminal coiled-coil participate in the tetrameric assembly of the channel; ATP binds at the nucleotide-binding domain and inhibits channel activity. TRPM4 has an exceptionally wide filter but is only permeable to monovalent cations such as sodium; filter residue Gln973 is essential in defining monovalent selectivity. The S1–S4 domain and the post-S6 TRP domain form the central gating apparatus that probably houses the Ca^{2+}- and PtdIns (4,5) P2-binding sites. These structures provide an essential starting point for elucidating the complex gating mechanisms of TRPM4 and reveal the molecular architecture of the TRPM family (Guo et al., 2017).

The TRPM4 channel is activated by an increase of intracellular Ca^{2+} and is regulated by several factors including temperature and P_i (4,5)P_z. TRPM4 allows Na^+ entry into the cell upon activation, but is completely impermeable to Ca^{2+}. TRPM4 proteins are widely expressed in the body. Currents with properties that are reminiscent of TRPM4 have been described in a variety of tissues since the advent of the patch clamp technology, but their physiological role is only beginning to be clarified with the increasing characterization of knockout mouse models for TRPM4. Furthermore, mutations in the *Trpm4* gene have been associated with cardiac conduction disorders in human patients (Marthar, et al., 2014).

The TRPM4 channel has been extensively studied in cerebral diseases such as stroke, head injury and multiple sclerosis. In the heart, gain-of-function mutations of TRPM4 are a cause of familial cardiac block and are involved in cardiac electrical activity, including diastolic depolarization from the sinoatrial node cells in mouse, rat, and rabbit, as well as action potential duration in mouse cardio-myocytes. In rat and mouse, pharmacological inhibition of TRPM4 prevents cardiac ischemia-reperfusion injuries and decreases the occurrence of arrhythmias. Several studies have identified TRPM4 mutations in patients with inherited cardiac diseases including conduction blocks and Brugada syndrome (Guinamard et al., 2015).

Recently, evidence has emerged to support the role of TRPM4 in certain types of cancer, such as prostate cancer and large B cell lymphoma.

The expression of TRPM4 could mediate certain behaviors of cancer cells such as migration and invasion. As a non-selective monovalent cation channel, TRPM4 upregulation and activation enhance sodium entry, which leads to depolarization of the membrane potential. The membrane potential is critical in regulating calcium influx, and a disturbed calcium homeostasis is always associated with cancer cell behaviors. Research on TRPM4 channels in cancer is at a very early stage; therefore, targeting the TRPM4 channel can be a novel way of managing cancer metastasis *via* disrupting calcium signaling pathways (Gao, Liao, 2019).

4.3.5. TRPM5

TRPM5 is a Ca^{2+}-activated cation channel that mediates signaling in taste and other chemosensory cells. Within taste cells, TRPM5 is the final element in a signaling cascade that starts with the activation of G protein-couple receptors (GPCR) by bitter, sweet, or umami taste molecules and that requires the enzyme PLCµ2. The latter breaks down PIP_2 into DAG and IP_3, and the ensuing release of Ca^{2+} from intracellular stores, activates TRPM5. Since its initial discovery in the taste system, TRPM5 has been found to be distributed in space chemosensory cells located throughout the digestive tract, in the respiratory system, and in the olfactory system. It is also found in pancreatic islets, where it contributes to insulin secretion (Liman, 2014).

TRPM5 is a chemo-sensor and its function in the gustatory and olfactory systems is well described. Previously, acute TRPM5 expression was reported in epithelial cells of the tongue. Global TRPM5 knockout (KO) mice present with impaired sweet taste response reduced body weight, fat mass gain and improved insulin resistance. While some metabolic aspects of this phenotype might indeed be due to non-functional TRPM5 channel in the tongue, caution is advised as these are global KO animals. In these mice, the *Trpm5* gene is disrupted in every cell and, therefore, the functional protein is never expressed. This is potentially problematic as compensatory mechanisms may develop due to the absence of the protein during development. In addition, the phenotype of these mice may be the compound effect of the absence of the channel in a variety of different cell populations in multiple tissues. To properly interpret a complex physiological phenotype such as reduced body weight, a complete mapping of TRPM5 expression throughout the body is needed (Wyatt et al., 2017).

It has been demonstrated that TRPM5 is expressed in the olfactory tissue of mice using a transgenic TRPM5-green fluorescent protein (GFP) line. Using the TRPM5-IC/eR26-tGFP strain, it was showed reporter gene expression in olfactory sensory neurons (OSNs) of the main olfactory epithelium (MOE), in microvilli and in the VNO (Pyrski et al., 2017). This study also revealed that *Trpm5-1*, the splice variant expressed in the taste buds, is acutely expressed in the adult MOE only in microvilli cells as these cells were co-labelled with antisera against TRPM5 (Wyatt et al., 2017).

The role of TRPM5 has been investigated with the use of *Trpm5* KO mouse models showing a lack of type II taste perception and a reduced glucose-induced insulin secretion. Expression levels of TRPM5 are reduced in obese, leptin-signalling-deficient mice, and mutations in TRPM5 have been associated with type II diabetes and metabolic syndrome (Vennekens et al., 2018). The TRPM5 channel is also likely to be involved in lung metastasis, as was shown by the increase in experimental lung metastasis when TRPM5 expression is enforced. TRPM5 activity increased the rate of acidic pH-induced MMP-9 expression, and required protein for the metastasis process. The use of triphenyl-phosphine oxide (TPPO), an inhibitor of TRPM5, in treatment of tumor-bearing mice, significantly reduced spontaneous lung metastasis (Hantute-Ghesquier et al., 2018; Maeda et al., 2017).

4.3.6. TRPM6

TRPM6 is a bifunctional protein comprising a TRP cation channel segment covalently linked to and α-type serin/threonine protein kinase.

TRPM6 is expressed in the intestinal and renal epithelial cells. Loss-of-function mutations in the human TRPM6 gene give rise to hypomagnesemia with secondary hypocalcemia (HSH), suggesting that the TRPM6 channel kinase plays a central role in systemic Mg^{2+} homeostasis. In contrast, TRPM6 null mice show a delay in prenatal development, neural tube detect, and prenatal death (Chubanov, Guderman, 2014). Unraveling the rare genetic diseases with manifestations of dysmagnesemia has greatly increased our understanding of the complex and intricate regulatory network in the kidney, specifically, functions of tight junction proteins including claudin-14, -16, -19, and -10; apical ion channels including: TRPM6, potassium channel ($K_v1.1$), and the renal outer medullary potassium channel (ROMK) (Li et al., 2017).

Magnesium is involved in many cell processes, including production of cellular energy, maintenance of DNA building blocks (nucleotides), protein production, and cell growth and death. Additionally, Mg^{2+} is needed for the production of a substance called parathyroid hormone that regulates blood calcium levels. Magnesium and calcium are also required for the normal functioning of nerve cells (motor neurons) that control muscle movement.

The TRPM6 channel is embedded in the membrane of epithelial cells that line the large intestine, structures in the kidneys known as distal convoluted tubules, the lungs, and the testes in males. When the body needs additional Mg^{2+}, the TRPM6 channel allows it to be absorbed in the intestine and filtered from the fluids that pass through the kidneys by the distal convoluted tubules. When the body has sufficient or too much Mg^{2+}, the TRPM6 channel does not filter out the Mg^{2+} from fluids but allows the ion to be released from the kidney cells into the urine. The channel also helps to regulate Ca^{2+}, but to a lesser degree (Chubanov, Guderman, 2014).

4.3.7. TRPM7

TRPM7 has been found in every mammalian tissue investigated to date. The two-in-one protein structure, the ubiquitous expression profile, and the proteins unique biophysical characteristics that enable divalent ion transport involve TRPM7 in a plethora of (patho) physiological processes. With its prominent role in cellular and systemic magnesium homeostasis, TRPM7 emerges as a key player in embryonic development, global ischemia, cardiovascular disease, and cancer (Fleig, Chubanov, 2014).

TRPM7 was initially proposed to regulate, and itself to be regulated by, intracellular Mg^{2+} levels, with Mg-ATP below 1mM strongly activating the channel. The importance of TRPM7 in cellular Mg^{2+} homeostasis has been investigated in several cell types, including leukocytes, platelets, vascular smooth muscle cells (VSMCs), cardiomyocytes, cardiac fibroblasts, osteoblasts, and tumor cells, in both physiological and pathological conditions (Zhou et al., 2019). Experiments using TRPM7 over-expressing cells demonstrated important mechanisms underlying the role of TRPM7 in Mg^{2+} homeostasis, and this was confirmed in the recently described crystal structure of mouse TRPM7, where partially hydrated Mg^{2+} ions occupy the center of the conduction pore (Duan et al., 2018a).

TRPM7-regulated Mg^{2+} has versatile biological functions, contributing to all vital cellular processes, including stability of tertiary structures of DNA and RNA, energy metabolism, enzyme activity, signaling, cell cycle progression and cell differentiation (Al Alawi et al., 2018). The significance of TRPM7 in development and cell viability was demonstrated in mice with global deletion of TRPM7. TRPM7 knockout mice are embryonic lethal and cardiac-targeted knockout of TRPM7 causes impaired embryonic development of the heart. The regulatory role of TRPM7 in cell differentiation has been highlighted in various cell types. In mesenchymal stromal cells, TRPM7 mediates shear stress and silencing TRPM7 accelerates osteogenic differentiation. TRPM7 was also shown to mediate differentiation in hepatic cells, lung fibroblasts, dental pulp stem cells and T cells (Ogunrinde et al., 2017; Zou et al, 2019).

Recently, several groups have identified small organic compounds acting as inhibitors or activators of the TRPM7 channel. In follow-up studies, the identified TRPM7 modulators were successfully used to uncover new cellular functions of TRPM7 *in situ* including a crucial role of

TRPM7 in Ca^{2+} signaling and Ca^{2+} dependent cellular processes. Hence, TRPM7 has been defined as a promising drug target (see review, Chubanov et al., 2017).

It has recently also reported that TRPM7 plays a significant role in ischemic and hypoxic brain injury and neuronal cell death. TRPM7, as a key non-glutamate mechanism of cerebral ischemia, also triggers an intracellular ionic imbalance and neuronal cell death in ischemia and hypoxia. This author has reported that TRPM7 is expressed in neurons of the hippocampus and cortex and activation of TRPM7 induced ischemic neuronal cell death; suppression of TRPM7 with virally mediated gene silencing using a mall interfering RNA (siRNA) reduced ischemic neuronal cell death and improved neurobehavioral outcomes *in vivo*. They also demonstrated that inhibition of TRPM7 using pharmacological means promoted neuronal outgrowth *in vitro* and provided neuroprotection against brain injury to hypoxia *in vivo* (Sun, 2017).

4.3.8. TRPM8

Transient receptor potential melastatin-8 (TRPM8) was originally cloned from prostate cancer tissue (Tsavaler et al., 2001). Shortly thereafter, the protein was identified as a cold- and menthol-activated ion channel in peripheral sensory neurons, where it plays a critical role in cold temperature detection. Functional and morphological studies have documented that TRPM8 is strongly expressed in a subpopulation of primary sensory neurons within the dorsal root and trigeminal ganglia. Thermo-sensitive nerve endings of these sensory neurons innervate the skin and mucosa (e.g., cornea, oral cavity, nasal epithelium). TRPM8 is a non-selective cation channel, with modest calcium permeability. Apart from physical (e.g., cold temperature) and chemical stimuli (e.g., menthol), TRPM8 is also gated by voltage. Altered TRPM8 expression or function has been linked to different disease states. Considering that TRPM8 shows a relatively restricted expression in normal tissues, this anomalous expression offers interesting possibilities as a diagnostic or prognostic marker and for therapeutic intervention in different types of cancer, in diseases of the urogenital tract, in cold pain and migraine, and of in airway and vascular pathologies (Almaraz et al., 2014).

Felix Viana team has recently found (Ordás et al., 2019) that TRPM8 is also expressed in the CNS. Although it is present at much lower levels than in peripheral sensory neurons, they found cells expressing TRPM8 in restricted areas of the brain, especially in the hypothalamus, septum, thalamic reticular nucleus, certain cortices and other limbic structures, as well as in some specific nuclei in the brainstem. Interestingly, positive fibers were also found travelling through the major limbic tracts, suggesting a role of TRPM8-expressing central neurons in multiple aspects of thermal regulation, including autonomic and behavioral thermoregulation. Additional *in situ* hybridization experiments in rat brain demonstrated a conserved pattern of expression of this ion channel between rodent species. These authors confirmed the functional activity of this channel in the mouse brain using electrophysiological patch-clamp recordings of septal neurons, and concluded that these results open a new window in TRPM8 physiology, guiding further efforts to understand potential roles of this molecular sensor within the brain (Ordás et al., 2019).

The TRP channels including TRPV1, TRPV4, TRPM8, and TRPA1, have been hypothesized to contribute to migraine, specifically as an activation mechanism of meningeal nociceptors. Reasons for the interest in a TRP channel contribution to migraine are largely due to their expression on meningeal nociceptors and their responsiveness to a variety of endogenous and exogenous stimuli that may be of relevance to migraine attacks. Furthermore, activation of TRP channels is well-known to promote the release of CGRP from sensory nerve endings, and they have been used extensively as probes for the function of CGRP in various processes. Given the role of CGRP in migraine described above, and the as yet unclear mechanism by which CGRP is released during migraine, TRP channels remain a focus of interest for their potential contribution to attacks. The TRPV1, TRPV4, TRPM8, and TRPA1 channels, based on their activation by pathological stimuli related to attacks or their modulation by drugs/natural products known to be efficacious for migraine. The repeated identification of TRPM8 variants in genome-wide association studies (GWAS) supports continued interest in this channel for the disorder, but it remains unknown whether therapeutics should be agonists or antagonists. Better understanding of the impact of the channel mutations on expression/function will help determine this answer (Benemei, Dussor, 2019; Geppetti et al., 2015).

In addition, TRPM8 channel plays a critical role in many pathophysiological processes. As stated above, TRPM8 was originally cloned in the prostate gland tissue (Tsavaler et al., 2001) and this channel has since then been identified as the cold/menthol receptor in the peripheral nervous system (McKemy et al., 2002). While this discovery led to hundreds of studies regarding the role of this channel in the peripheral nervous system and the associated pain and thermo-sensation phenomena (see review Señaris et al., 2018), the involvement of TRPM8 in cancer was relative neglect. Despite this fact, there is growing evidence that TRPM8 should be carefully studied within the frame of carcinogenesis, especially in the prostate, where it is highly expressed and where many teams have confirmed variations in its expression during cancer progression. Among many other points that still need to be addressed the question of endogenous modulators of TRPM8 in the prostate and of their mechanisms of action is a critical one. Evidence of regulation by physiological factors such as prostate-specific antigen (PSA), androgens, and *lipoprotein lipase* (LPL) has proved that TRPM8 can exhibit an activity beyond that of cold receptor, thus explaining how the channel can be activated in organs not exposed to temperature variations (Noyer et al., 2018).

4.4. TRPML Subfamily

The TRPML subfamily consists of three members: the founding member TRPML1, TRPML2 and TRPML3. TRPML1 was identified as the protein mutated in the lysosomal storage disease, mucolipidosis type IV (MLIV). TRPML2 was found by database searches and TRPML3 was identified as the protein mutated in mice with the varitint-waddler phenotype. The latter is characterized by auditory, vestibular, and melanocytes phenotype, which is caused by the gain-of-function mutation A419P in the predicted fifth transmembrane domain of TRPML3 (Yamaguchi, Muallem, 2010). The functions of TRPML proteins include roles in vesicular trafficking and biogenesis, maintenance of neuronal development, function, and viability, and regulation of intracellular and organelle ionic homeostasis (Venkatachalam et al., 2015).

4.4.1. TRPML1

The first member of the mammalian mucolipin TRP channel subfamily (TRPML1 or MCOLN1) is a cation-permeable channel that is predominantly localized on the membranes of late endosomes and lysosomes (LELs) in all mammalian cell types. In response to the regulatory changes of LEL-specific phosphoinositides or other cellular cues, TRPML1 may mediate the release of Ca^{2+} and heavy metal Fe^{2+}/Zn^{2+} ions into the cytosol from the LEL lumen, which in turn may regulate membrane trafficking events (fission and fusion), signal transduction, and ionic homeostasis in LELs. Human mutation in TRPML1 results in the type IV mucolipidosis (ML-IV), a childhood neurodegenerative lysosomes storage disease. At the cellular level, loss-of-function mutations of mammalian TRPML1 or its *C. elegans* or *D. melanogaster* homolog gene results in lysosomal trafficking defects and lysosome storage (Wang et al., 2014; Waller-Evans, Lloyd-Evans, 2015).

Mutations in the gene encoding TRPML1 cause MLIV, a recessive lysosomal storage disorder (LSD) characterized by neurodegeneration, psychomotor retardation, ophthalmologic defects and achlorhydria. The MLIV patient fibroblasts show enlargement and engulfment of the late endo-lysosomal compartment, autophagy impairment, and accumulation of lipids and glycosaminoglycans (see review Di Paola et al., 2018).

4.4.2. TRPML2

The TRPML2 protein, encoded by the *Mcoln2* gene, is one of the three mucolipins (TRPML1-3), a subset of the TRP superfamily of ion channels. Although there are no thorough studies on the cellular distribution of TRPML2, its mRNA appears to be largely restricted to lymphocytes and other immune cells. This contrasts with the ubiquitous expression of TRPML1 and the limited but divers expression of TRPML3 clearly suggests a specialized role for TRPML2 in immunity. Localization studies indicate that TRPML2 is present in lysosomes (including the specialized lysosome-related organelle that B-lymphocytes use for processing of the antigen – bound B-cell receptor), late endosomes, recycling endosomes, and, at a much lower level, the plasma membrane. Heterologously expressed TRPML2, like TRPML1 and/or TRPML3, forms ion channels that can be activated by a gain-of-function mutation (alanine to proline in the fifth transmembrane domain, close to the pore) that favors the open state, by a transient reduction of extracellular sodium followed by sodium replenishment, by small chemicals related to sulfonamides, and by PI(3,5)P2, a rare phosphoinositide that naturally accumulates in the membranes of endosomes and lysosomes and this could act as a physiologically relevant agonist. TRPML2 channels are in virally rectifying and permeable to Ca^{2+}, Na^+, and Fe^{2+}. When heterologously co-expressed, TRPML2 can form heteromultimers with TRPML1 and TRPML3. In B-lymphocytes, TRPML2 and TRPML1 may play redundant roles in the function of their specialized lysosome. Although the specific subcellular function of TRPML2 is unknown, distribution and channel properties suggest roles in calcium release from endolysosomes, perhaps to regulate vesicle fusion and/or subsequent session or to release calcium from intracellular acidic stories for signaling in the cytosol. Alternatively, TRPML2 could function in the plasma membrane, and its abundance in vesicles of the endocytic pathway could simply be due to regulation by endocytosis and exocytosis. Evolutionary point of view, vertebrate mucolipins arise in the period, in which adaptive immunity appeared. Given the restricted expression of TRPML2 in immune cells, this evolutionary history suggests a functional role the adaptive immunity characteristic of vertebrates (García-Añoveros, Wiwatpanit, 2014).

As stated above, TRPML1 is associated with the human lysosomal storage disease, mucolipidosis (MLIV), but TRPML2 and TRPML3 have not been linked with a human disease. Although TRPML1 is expressed in many tissues, TRPML3 is expressed in a varied but limited set of tissues, while TRPML2 has a more limited expression pattern where it is mostly detected in lymphoid and myeloid tissues. TRPML2 is essential in endosome-lysosome interaction, the recycling of GPI-APs, and possibly the trafficking of immune-associated proteins such as MHC-I, TLRs, and Fc-εRI to name a few. In addition to its possible role in adaptive immunity through proper B-cell differentiation and maturation, TRPML2 may also be involved in type I hypersensitivity. While many aspects of the function of TRPML2 remain to be revealed, future research will undoubtedly shed light on its importance in cells and perhaps even unlock its potential to complement for the functional loss of TRPML1 protein in MLIV disease (Cuajungco et al., 2016).

4.4.3. TRPML3

TRPML3 belongs to the TRPML (MCOLN) subfamily of TRP channels comprising three genes in mammals. The TRPML3 protein features a six-transmembrane topology and functions as a Ca^{2+} permeable inward rectifying cation channel that is open at sub-physiological pH and closes as the extra-cytosolic pH becomes more acidic. TRPML3 localizes to the plasma membrane and to early- and late-endosomes as well as lysosomes (Noben-Trauth, 2011).

Gain-of-function mutations in TRPML3 channel cause deafness and circling behavior in mice. A special feature of TRPML3 is their intracellular expression. These channels mostly reside in membranes of organelles of the endosomal system such as early and late endosomes, recycling endosomes, lysosomes, or lysosome–related organelles. Although the physiological roles of TRPML subfamily channels within the endo-lysosomal system are far from being fully understood, it is speculated that they are involved in the regulation of endo-lysosomal pH, fusion/fission processes, trafficking, autophagy, and/or secretion (e.i., hormones) and exocytosis (Grimm et al., 2014).

Endolysosomal dysfunction can cause lysosomal storage disorders such as mucolipidoses, sphingolipidoses, or neuronal ceroid lipofuscinoses, which typically go along with fatal neurodegenerative processes. Dysfunction of the endolysosomal system has also been implicated in the development of many other diseases, ranging from metabolic diseases, retinal and pigmentation disorders to infectious diseases and even cancer. TRPML3 is shown to be functionally active in both early endosomes and late endosomes/lysosomes in over-expressing cells as well as in endogenously expressing CD11b+ lung-tissue macrophages (Chen et al., 2017).

It has just recently been established that the endolysosomal, non-selective cation channels, two-pore channels (TPCs) and mucolipins (TRPMLs) as TPC/TRPML interaction partners ('interactomes') regulate intracellular membrane dynamics and autophagy. While partially compensatory for each other, isoform-specific intracellular distribution, cell-type expression patterns, and regulatory mechanisms suggest different channel isoforms confer distinct properties to the cell. In recently published review article of Krogsaeter and colleagues (2019) TRPML3 interactoms are shown, and a meta-analysis of experimentally obtained channel interactomes conducted by authors. Accordingly, interactomes are compared and contrasted, and subsequently described in detail for TPC1, TPC2, TRPML1, and TRPML3. The latter channel protein appears up-regulated in squamous cell carcinoma and hepato-cellular carcinoma, and down-regulated in erythro-leukemia and platelets of cancer patients. The potential implications of TRPML3 in immunological and oncogenic signaling pathways may be of particular relevance. The authors underscore, thus, the potential held by TPC/TRPML as future therapeutic targets to treat currently incurable diseases, while identifying approaches necessary to elucidate the multifaceted roles of endo-lysosomal, non-selective cation channels (Krogsaeter et al., 2019).

4.5. TRPN Subfamily

Mechano-sensory transduction for senses such as proprioception, touch, balance, acceleration, hearing and pain relies on mechano-transduction channels, which convert mechanical stimuli into electrical signals in specialized sensory cells. How force gates mechano-transduction channels is a central question in the field, for which there are two major models. One is the membrane tension model: force applied to the membrane generates a change in membrane tension that is sufficient to gate the channel, as in the bacterial large conductance *mechanosensitive ion channel* (MscL) and certain eukaryotic potassium channels. The other is the tether model: force is transmitted *via* a tether to gate the channel (Jin et al., 2017).

The 'no mechanoreceptor potential C' (NOMPC) or TRPN channel is a diverse group of proteins thought to be involved in mechanoreception and is important for mechanosensation-related behaviors such as locomotion, touch and sound sensation across different species including *Caenorhabditis elegans*, *Drosophila* and zebrafish. NOMPC is the founding member of the TRPN subfamily, and is thought to be gated by tethering of its ankyrin repeat domain to microtubules of the cytoskeleton. TRPN fulfils all the criteria that apply to mechanotransduction channels and has 29 ankyrin repeats, the largest number among TRP channels. A key question is how the long ankyrin repeat domain is organized as a tether that can trigger channel gating. A single-particle ECM structural analysis suggests that the ankyrin repeat domain of TRPN channel resembles a helical spring, suggesting its role of linking mechanical displacement of the cytoskeleton to the opening of the channel pore. The TRPN architecture underscores the basis of translating mechanical force into an electrical signal within a cell (Jin et al., 2017).

Recent behavioral observation revealed that the locomotion of the double-stranded RNA (dsRNA) injected nymphs of the insect *Nilaparvata lugens* was defective with much less movement compared to the negative control. Feeding and honeydew excretion of the dsRNA injected insects also decreased significantly. They cloned the *NompC* gene of N. lugens and found that the full length *NompC* of N. lugens possessed similar structure as *Drosophila NompC*, which belongs to TRPN subfamily. The expression pattern analysis of different developmental stages and body parts showed that the transcription of *NompC* of N. lugens was more abundant in adult stage and in the abdomen. Injection of dsRNA of N. lugens *NompC* in the third-instar nymphs successfully knocked down the target gene with 75% suppression. At nine days after injection, the survival rate of dsRNA injected nymphs was as low as 10%. These results suggested that N. lugens *NompC* is a classical mechano-transduction channel that plays important roles in proprioception and locomotion, and is essential for the survival of this insect (Wang et al., 2019).

4.6. TRPP Subfamily

It has been exciting times since the identification of polycystic kidney disease 1 (PKD1) and PKD2 as the genes mutated in autosomal dominant polycystic kidney disease (ADPKD). Biological roles of the encoded proteins TRPP1 (Polycystin-1) and TRPP2 have been deduced from phenotypes in ADPKD patients, but recent insights from vertebrate and invertebrate model organisms have significantly expanded our understanding of the physiological functions of these proteins. The identification of additional TRPP3 and TRPP5 channels and polycystin-1-like proteins (PKD1L1, PKD1L2, PKD1L3, and PKDREJ) has added yet another layer of complexity to these fascinating cellular signaling units. TRPP proteins assemble with polycystin-1 family members to form receptor–channel complexes. These protein modules have important biological roles ranging from tubular morphogenesis to determination of left-right asymmetry. The founding members of the polycystin family, TRPP1 and TRPP2-1, are a prime example of how studding human disease genes can provide insights into fundamental biological mechanisms using a so-called 'reverse translational' approach (Semmo et al., 2014).

Structural point of view, recently David Clapham group has reported that a structure of the full-length human polycystin 2-11 protein conforms to the TRPP polycystin-2 structures and established the core structure and point out differences in regions and residues that may account for polycystin-2 and polycystin 2-l1's distinct permeation and gating (Hulse et al., 2018). It should be noted here that members of the TRPP and TRPML subfamilies share a large, globular-folded extracellular/luminal domain named tetragonal opening for polycystins (TOP) domain in TRPP and polycystin-mucolipin domain (PMD) in TRPML. The structures of TRPP2, TRPML1 and TRPML3 revealed intriguing similarities with respect to Ca^{2+} binding and activation by negatively charged lipids. TRPML1 is located within endosomal and lysosomal membranes and directly involved in type IV mucolipidosis, while TRPP2 is located mainly in the endoplasmic reticulum (ER) and depending on cell type in the primary cilium and the plasma membrane. Despite these similarities, a striking difference is observed. In particular, the coordination of the PMD and the TOP domain within the tetrameric assembly is substantially altered. In TRPP2, the unique $β$-hairpin with the three-leaf clover fold interacts with S3a helix-extensions, likely transducing the conformational changes directly into the membrane-integral part of the channel. The different settings of open/closed states found in TRPML and TRPP2 structures might provide some insight into the puzzling localization-depended functional variations (Madej, Ziegler, 2018).

4.7. TRPV Subfamily

Within the TRP superfamily of ion channels, the vanilloid (TRPV) subfamily came into existence and immediately gained public notoriety in 1997, when its founding members, TRPV1 in mammals and OSM-9 in *C. elegans*, were first reported (Caterina et al., 1997; Colbert et al., 1997). TRPV1 was identified by an expression cloning strategy. TRPV2, TRPV3, and TRPV4 were identified by a candidate gene approach, respectively. The TRPV channels can be sub-grouped into four branches by sequence comparison. One branch includes four members of mammalian TRPVs (TRPV1, TRPV2, TRPV3, RPV4); *in vitro* whole-cell recording showed that they respond to temperatures higher than 42, 52, 31, and 27°C, respectively, suggesting that they are involved in thermosensation, hence the term 'thermo-TRPs' (Caterina, 2007; Moore, Liedtke, 2017). The second mammalian branch includes the Ca^{2+}-selective channels, TRPV5 and TRPV6, possibly subserving Ca^{2+} uptake in the kidney and intestine. One invertebrate branch includes *C. elegans* OSM-9 and *Drosophila* Inactive (IAV) *trpv* gene; the other branch comprises OCR-1 to OCR-4 in *C. elegans* and *Drosophila trpv* Nanchung (NAN) gene (Moore, Liedtke, 2017).

Structural point of view, the TRPV subfamily consists of six homologous members with diverse functions. TRPV1–TRPV4 members are nonselective cation channels proposed to play a role in nociception, while TRPV5 and TRPV6 are involved in epithelial Ca^{2+} homeostasis. The cryo-EM map suggests that TRPV subfamily members have highly homologous structural topologies. These results allowed postulating a structural explanation for the functional diversity among TRPV channels and their differential regulation by proteins and ligands (Huynh et al., 2014).

4.7.1. TRPV1

TRPV1 is a well-characterized channel expressed by a subset of peripheral sensory neurons involved in pain sensation and also at a number of other neuronal and non-neuronal sites in the mammalian body. Functionally, TRPV1 acts as a sensor for noxious heat (greater than 42°C). It can also be activated by some endogenous lipid derived molecules, acidic solution (pH < 6.5) and some pungent chemical and food ingredients such as capsaicin, as well as by toxins such as resiniferatoxin (RTX) and vanillotoxins (see for review, Jeon, Caterina, 2018). Structurally, TRPV1 subunits have six trans-membrane (TM) domains with intracellular N- (containing 6 ankyrin-like repeats) and C-termini, and a pore region between TM5 and TM6 containing sites that are important for channel activation and ion selectivity. The N- and C-termini have residues and regions that are sites for phosphorylation/dephosphorylation and PI (4.5) P2 binding, with regulation TRPV1 sensitivity and membrane insertion. The channel has several interacting proteins, some of which (e.g., AKAP7a/150) are important for TRPV1 phosphorilation. Four TRPV1 subunits from a non-selective, outwardly rectifying ion channel permeable to monovalent and divalent cations with a single-channel conductance of 50-100 pS. TRPV1 channel kinetics reveals multiple open and closed states, and several models for channel activation by voltage ligand binding and temperature have been proposal. Studies with TRPV1 agonists and antagonists and *Trpv1* mice have suggested a role for TRPV1 in pain, thermoregulation and osmoregulation, as well as in cough and overactive bladder. TRPV1 antagonists have advanced to clinical trials where findings of drug-induced hyperthermia and loss of heat sensitivity have raised question about the viability of than therapeutic approach (Bevan et al., 2014). However, few orally administered TRPV1 antagonists are still being considered for clinical use.

Somewhat paradoxically, TRPV1 agonists also have analgesic properties. Dilute (0.1%) capsaicin cream is available as an over-the-counter pain reliever. A more concentrated (8%) capsaicin patch, marketed as Qutenza, is approved for the treatment of postherpetic neuralgia in the United States. Whereas moderate doses of capsaicin can lead to reversible analgesia, intrathecal or intraganglionic injection of the ultrapotent TRPV1 activator RTX has been used to create permanent analgesia by destroying the signaling potential of susceptible sensory neurons. Dramatic reductions in pain behaviors after intrathecal treatment with RTX have been demonstrated in multiple species, including dogs suffering from debilitating pain due to osteosarcoma. Collectively, these data support the exciting clinical trials under way exploring the benefits of RTX in patients with intractable cancer pain (Moran, 2018).

4.7.2. TRPV2

TRPV2 is a calcium-permeable cation channel belonging to the TRPV family. This channel is activated by heat more > 52°C, various ligands, and mechanical stresses. In most of the cells, a large portion of TRPV2 is located in the endoplasmatic reticulum under unstimulated conditions. Upon stimulation of the cells with PI-3-kinase-activating ligands, TRPV2 is translocated to the plasma membrane and functions as a cation channel. Mechanical stress may also induce translocation of TRPV2 to the plasma membrane. The expression of TRPV2 is high in same types of cells including neurons, neuro-endocrine cells, immune cells involved in innate immunity, and certain types of cancer cells. In turn TRPV2 may modulate various cellular functions in these cells (Kojima, Nagasava, 2014). While TRPV2 was initially characterized as a noxious heat sensor, later it was found that TRPV2 can also act as a mechanosensor in embryonic neurons or adult myenteric neurons. Combined, these results demonstrate that the same channel can have many distinct functions depending on its location. TRPV2 can also act as a lipid sensor; therefore further identification of novel physiological roles for TRPV2 will be dependent on its pattern of expression. Recently, some reports have indicated that TRPV2 is involved in the disease progression of bladder or prostate cancer. It was found that there was a significant relationship between the over-expression of TRP genes, including TRPV2, and the survival of patients with glioblastoma. These results demonstrate that TRP channels contribute to the progression and survival of glioblastoma patients (Shibasaki, 2016a).

It has just recently confirmed that TRPV2 is expressed in both neural progenitor cells and glioblastoma stem/progenitor-like cells (GSCs). In developing neurons, post-translational modifications of TRPV2 (e.g., phosphorylation by ERK2) are required to stimulate Ca^{2+} signaling and nerve growth factor (NGF)–mediated neurite outgrowth. TRPV2 over-expression also promotes GSC differentiation and reduces glioma genesis *in vitro* and *in vivo*. In glioblastoma, TRPV2 inhibits survival and proliferation, and induces Fas/CD95-dependent apoptosis. Furthermore, by proteomic analysis, the identification of a TRPV2 interactome-based signature and its relation to glioblastoma progression/recurrence, high or low overall survival and drug resistance strongly suggest an important role of the TRPV2 channel as a potential biomarker in glioblastoma prognosis and therapy (Santoni, Amantini, 2019).

4.7.3. TRPV3

TRPV3 is the Ca^{2+}-permeable nonselective cation channel widely expressed in skin keratinocytes as well as oral and nasal epithelia. TRPV3 is activated by innocuous warm as well as noxious hot temperature. Activation of TRPV3 in skin keratinocytes causes release of multiple substances, which in turn regulate diverse functions including skin barrier formation, hair growth, wound healing, temperature sensing, and itch and pain perceptions. While several natural and synthetic ligands have been described for TRPV3, only one of them, farnesyl pyrophosphate (FPP), is naturally produced in animal cells. Together wide the use of genetic mouse models, application of these compounds have revealed not only the physiological functions but also regulatory mechanisms of TRPV3 channel by extracellular Ca^{2+}, Mg^{2+} and protons as well as intracellular Ca^{2+}-calmodulin, ATP, PI-4,5- bisphospate, polyunsaturated fatty acids, and Mg^{2+}. Gain-of-functions genetic mutations of TRPV3 in rodents and humans have been instrumental in unveiling the critical role of this channel in skin health and disease. Although being less studied as compared as to its closed relatives, TRPV1 and TRPV3 have turned out to be a very important channel for skin health. A property balanced function of TRPV3 appears to be critical for skin barrier formation, hair growth, wound healing, keratinocyte mutation, and cutaneous pain, itch, and temperature sensation. Naturally, occurring genetic mutations with augmented TRPV3 function they lead to hair less, skin inflammation, severe itchiness and dermatitis, suggesting a potential use of TRPV3 blockers in the treatment of skin disease (Yang, Zhu, 2014).

The cryo-EM structures of the human TRPV3 channel in multiple conformational states provide a glimpse of putative gating intermediates.

The apo-structure of TRPV3, despite being closely related to other thermo-TRPV channels, shows a number of unique features. Most notably, in contrast to the selectivity filters (SFs) of TRPV1 and TRPV2, which are occluded by side chains of hydrophobic residues, the SF of TRPV3 is lined with backbone carbonyl groups and is wide enough to permeate partially dehydrated cations. Interestingly, the SF remains mostly unchanged in all the conformational states captured, which suggests that the SF in TRPV3 does not act as a gate, in contrast to TRPV1 and TRPV2 (Zubcevic et al., 2018).

TRPV3 is activated by innocuous temperature, with the threshold for channel opening at 31–39°C, and activity is retained over temperatures extending into the noxious range. Chemical agonists include spice extracts (such as camphor and carvacrol), synthetic agents (including 2-aminoethoxy diphenylborate, 2-APB) and endogenous ligand FPP. Repeated exposure to heat or chemical agonists leads to sensitization of the receptor, through hysteresis of gating, while co-application of diverse stimuli, such as heat and chemical is synergistic. To date information from preclinical data utilizing structurally diverse TRPV3 antagonists do provide some support that blockade of this channel may provide analgesia in chronic pain states (Broad et al., 2016).

With respect to skin physiology and pathophysiology, probably TRPV3 is the most important TRPV channel (Figure 11). Cutaneous keratinocytes are one of the key cell types of the skin orchestrating a plethora of functions. No wonder therefore that TRPV3 which is highly expressed by these cells is suggested to be involved in numerous cutaneous regulatory mechanisms and phenomena such as the formation and maintenance of the physical–chemical skin barrier, hair growth, growth and survival of inter-follicular skin cell populations, cutaneous inflammation and pain, and pruriceptive itch as well (Nilius, Bíró, 2014).

The introduced experimental data as well as the mutant human and mouse phenotypes strongly suggest that TRPV3 antagonists might be useful tools in the management of a wide array of skin diseases, e.g., some dermatoses in which the beneficial effects of inhibiting TRPV3 activity could be predicted (Nilius, Bíró, 2014). For example, multiple 'gain-of-function' mutations of TRPV3 were identified in Olmsted syndrome, the first cutaneous TRP channelopathies. This rare geno-dermatosis is characterized by the development of hyper-orthokeratosis and keratomas, diffuse alopecia and extreme pruritus (Nilius ete al., 2014).

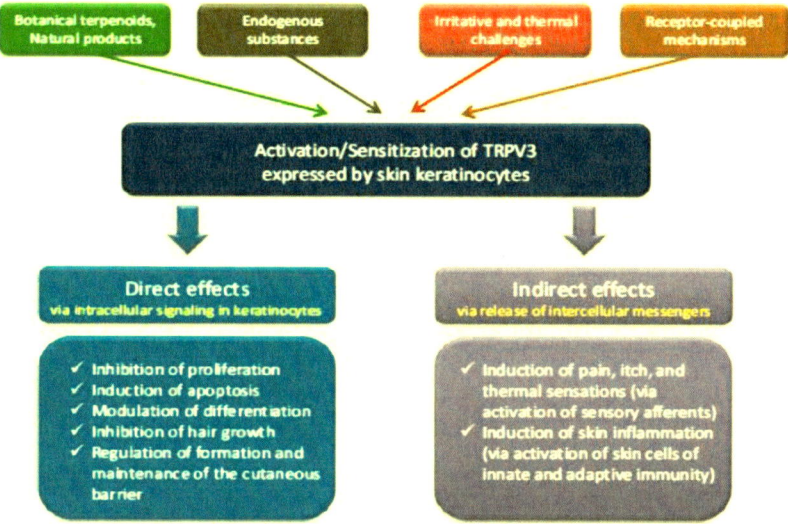

Figure 11. The central role of TRPV3 in the regulation of multiple cutaneous functions. TRPV3, expressed by epidermal and hair follicle keratinocytes, can be activated and/or sensitized by various exogenous agents (e.g., botanical terpenoids, a hydroxyl acids), endogenous substances [e.g., 17(R)-resolvin D1, FPP], sensory heat stimuli and intracellular signaling pathways. These stimuli induce a TRPV3- mediated influx of Ca^{2+} to the keratinocytes which, by initiating mostly unidentified downstream mechanisms, results in dual effects, (i) as direct effects, and (ii) indirect effects (reproduced from Nilius, Bíró, 2014 with permission).

4.7.4. TRPV4

TRPV4 is a Ca^{2+}-permeable nonselective cation channel that is activated by a disparate array of stimuli, ranging from hypotonicity to heat and acid. This widely distributed TRPV4 channel participates in the transduction of both physical (osmotic, mechanical, and heat) and chemical (endogenous, plant-derived, and synthetic ligands) stimuli. Compared to the vast knowledge obtained about TRPV4 channel regulation, little is known about the control of TRPV4 transcription. Progesterone receptor mediates repression of TRPV4 transcription in epithelial and vascular smooth muscle cells. Down-regulation of TRPV4 expression by micro-RNA 203 in condylar cartilage of the temporo-mandibular joint and by probiotic bacteria strains in the colon has also been reported. Inflammatory signals such as interleukin 1β and interleukin 17 increase TRPV4 mRNA levels in dorsal root ganglia (DRG) neurons, and nerve growth factor (NGF) increases TRPV4 expression in the urothelium. Hypoxia/ischemia increases TRPV4 expression and function in astrocytes and in pulmonary arterial smooth muscle cells of mice exposed to chronic hypoxia-induced pulmonary hypertension (see for review, Garcia-Elias et al., 2014; White et al., 2016).

TRPV4 is now recognized as being a polymodal ionotropic receptor that is activated by a disparate array of stimuli, ranging from hypotonicity to heat and acidic pH. Importantly, this ion channel is constitutively expressed and capable of spontaneous activity in the absence of agonist stimulation, which suggests that it serves important physiological functions, as does its widespread dissemination throughout the body and its capacity to interact with other proteins. It has emerged more recently that TRPV4 fulfills a great number of important physiological roles and that various disease states are attributable to the absence, or abnormal functioning of this ion channel (White et al., 2016).

Recently TRPV4 has been identified as an ion channel that is modulated by, and opened by intracellular signaling cascades from other receptors and signaling pathways. Although TRPV4 knockout mice show relatively mild phenotypes, some mutations in TRPV4 cause severe developmental abnormalities, such as the skeletal dysplasia and arthropathy. Regulated TRPV4 function is also essential for healthy cardiovascular system function as a potent agonist compromises endothelial cell function, leading to vascular collapse. A better understanding of the signaling mechanisms that modulate TRPV4 function is necessary to understand its physiological roles. Post-translational modification of TRPV4 by kinases and other signaling molecules can modulate TRPV4 opening in response to stimuli such as mechanical and hypo-osmolarity and there is an emerging area of research implicating TRPV4 as a transducer of these signals as opposed to a direct sensor of the stimuli. Due to its wide expression profile, TRPV4 is implicated in multiple pathophysiological states. TRPV4 contributes to the sensation of pain due to hypo-osmotic stimuli and inflammatory mechanical hyperalgesia, where TRPV4 sensitization by intracellular signaling leads to pain behaviors in mice. In the vasculature, TRPV4 is a regulator of vessel tone and is implicated in hypertension and diabetes due to endothelial dysfunction. TRPV4 is a key regulator of epithelial and endothelial barrier function and signaling to and opening of TRPV4 can disrupt these critical protective barriers. In respiratory function, TRPV4 is involved in cystic fibrosis, cilary beat frequency, broncho-constriction, chronic obstructive pulmonary disease, pulmonary hypertension, acute lung injury, acute respiratory distress syndrome and cough (Darby et al., 2016).

In addition to the unique nociceptor or pluriceptor functions of TRPV4, this channel is an important regulator in brain activity. Mutations of TRPV4 relate to many severe diseases such as scapula-peroneal spinal muscular atrophy, Charcot-Marie-Tooth disease type 2C, and skeletal dysplasias. Especially, most TRPV4 mutations cause peripheral neuropathy. These results strongly indicate that TRPV4 is also an important molecular target for cure of neuropathy (Shibasaki, 2016b). Recent results from animal studies suggest that TRPV4 antagonism has therapeutic potential in edema, pain, gastrointestinal disorders, and lung diseases. A lack of observed side-effects *in vivo* has prompted a first-in-human trial for a TRPV4 antagonist in healthy participants and stable heart failure patients (Grace et al., 2017).

Finally, a growing body of studies showed that TRPV4 acted as a crucial regulator in the progression of fibrosis including myocardial fibrosis, cystic fibrosis, pulmonary fibrosis, hepatic fibrosis and pancreatic fibrosis, suggesting TRPV4 may be a potential therapeutic vehicle in fibrotic diseases. In particular, increasing evidence shows that TRPV4 modulated fibroblasts proliferation and differentiation to myofibroblasts by integrating mechanical and soluble signals that are derived from extracellular matrix (ECM) stiffness and *transforming growth factor beta* 1 (TGF-β1). The regulation of obesity by TRPV4 seems to be a novel mechanism underlying TRPV4 contribute to fibrosis (Zhan, Li, 2018).

4.7.5. TRPV5

TRPV5 is one of the two channels in the TRPV family that exhibit high selectively to Ca^{2+} ions. TRPV5 mediates Ca^{2+} reflux in to cells as the first step to transport Ca^{2+} across epithelia. The specialized distribution in the distal tubule of the kidney positions TRPV5 is a key player in Ca^{2+} reabsorption. The responsiveness in expression and/or activity of TRPV5 to hormones such as 1,25-dihidroxyvitamin D_3, parathyroid hormone, estrogen, and testosterone wakes TRPV5 suitable for its role in the fine-tuning of Ca^{2+} reabsorption. This role is further optimized by the modulation of TRPV5 trafficking and activity *via* its binding partners; co-expressed proteins, tubular factors such as carbindin-D_{28k}, calmodulin, transmembrane protein *klotho*, uromodulin, and plasmin; extracellular and intracellular factors such as H^+, Mg^{2+}, Ca^{2+}, PI-4,5-bisphospate, and fluid flow. This regulation allows TRPV5 to adjust its overall activity in response to the body's demand for Ca^{2+}, and to prevent kidney stone formation. A point mutation in mouse *Trpv5* gene leads to hyper-calciuria similar to *Trpv5* knockout mice, suggesting a possible role of TRPV5 in hypercalciuric disorders in humans. In addition, the single nucleotide polymorphisms in *Trpv5* gene, prevalently present in African descents, may contribute to the efficient renal Ca^{2+} reabsorption among African descendants. TRPV5 represents a potential therapeutic target for disorders with altered Ca^{2+} homeostasis (Na, Peng, 2014).

It is known that two TRPV subfamily channels, TRPV5 and TRPV6, have evolved into highly selective ion channels that partake in the process of epithelial calcium transport, contributing significantly to the maintenance of human calcium homeostasis. Transcellular calcium active transport is a regulated process involving three steps; apical entry of calcium into epithelial cells *via* TRPV5 or TRPV6, immediate binding of calcium by calbindin-D_{9k} and calbindin-D_{28K} which transport calcium to the basolateral membrane, and extrusion through the plasma membrane calcium ATPases (PMCA1 and PMCA4) and sodium-calcium exchanger isoform 1. TRPV5 and TRPV6 are co-expressed in intestinal and renal epithelia and are able to form hetero-tetrameric channels. TRPV5 is the main isoform in renal epithelia, whereas TRPV6 is the main isoform in the intestines. Of note, TRPV5 expression is restricted to kidney while TRPV6 is more ubiquitously expressed (van Goor et al., 2017).

It was discovered early on that TRPV5 and TRPV6 channels exhibit a rundown of activity during prolonged electrophysiological recordings. To date, however, most patch-clamp studies of TRPV5/6 have been performed in heterologous expression systems. It should be kept in mind that in the conventional whole-cell recording configuration used for TRPV5/6 recording the ER stores will also most likely be depleted due to low calcium and the presence of chelators in the internal solution. A way around this would be to use a perforated patch, which however will prevent the depletion of cytoplasmic calcium and reduce current amplitudes by inactivation (Kozak, Rychkov, 2018).

A major difference between the properties of TRPV5 and TRPV6 lies in their tissue distribution: TRPV5 is predominantly expressed in the distal convoluted tubules and connecting tubules of the kidney, with limited expression in extra-renal tissues. In contrast, TRPV6 exhibits a broader expression pattern, showing prominent expression in the intestine with additional expression in the kidney, placenta, epididymis, exocrine tissues (i.e., pancreas, prostate, salivary gland, sweat gland), and a few other tissues. Thus, while TRPV5 plays a key role in determining the level of urinary Ca^{2+} excretion, the physiological roles of TRPV6 are not limited to intestinal Ca^{2+} absorption (Peng et al., 2018).

4.7.6. TRPV6

The TRPV6 channel was first isolated from rat small intestine followed by the identification of the human and mouse orthologues.

TRPV6 displays several specific features which makes it unique among the members of the mammalian *TrpV* gene subfamily: (1) TRPV6 (and its closest relative, TRPV5) are the only highly Ca^{2+}-selective channels of the entire TRP superfamily; (2) Translation of *Trpv6* initiates at a non-AUG codon, at ACG, located upstream of the annotated AUG, which is not used for initiation. The ACG codon is nevertheless decoded by methionine. Not only is a very rare event in eukaryotic biology, the full-length TRPV6 protein existing *in vivo* comprises an amino terminus extended by 40 amino acid residues compared to the annotated truncated TRPV6 protein which has been used in most studies on TRPV6 channel activity so far; (3) Only in humans a coupled polymorphism of *Trpv6* exists causing three amino acid exchanges and resulting in an ancestral *Trpv6* haplotype and a so-called derived *Trpv6* haplotype. The ancestral haplotype is found in all species, while the derived *Trpv6* haplotype has only been identified in humans, and its frequency increases with the distance to the African continent. Apparently the *Trpv6* gene has been a strong target for selection in humans, and its derived variant is one of the few examples showing consistently differences to the orthologues genes of other primates; (4) The *Trpv6* gene expression is significantly upregulated in several human malignancies including the most common cancers, prostate and breast cancer; (5) Male mice lacking functional TRPV6 channels are hypo-/infertile making TRPV6 one of the very few channels essential for male fertility (Fecher-Trost et al., 2014, 2017).

Concerning the TRPV6 in cancer research, significant advances have been made during last two decades as to the discovery of TRPV6 channel expression in various cancer tissues. Among them are the cancers of the epithelial origin such as of prostate, breast, pancreas, ovaries, endometrium, testicule, colon, and lung. Though its role in cancer cell survival, proliferation, and apoptosis resistance was already established both *in vitro* and *in vivo*, much less was done in the studying of the downstream pathways where TRPV6 channel is involved which are restricted so far to the Ca^{2+} – calmodulin – calcineurin – NFaT (calcium-dependent nuclear factors in activated T cells) pathways. Despite the discovery of the crucial role of TRPV6 in cancer cell proliferation and survival *in vitro*, no reliable tool to target TRPV6 channel *in vivo* has been reported so far to be used as effective therapy against above cancers. TRPV6 does represent a prospective target in cancer treatment because of its involvement in cancer cell proliferation, metastasis development and apoptosis inhibition. A number of TRPV6 inhibitors are already known and are mentioned in targeting calcium signaling in cancer. However, only some of them are specific and discriminative for the TRPV6 channel (e.g., ruthenium red, 2-aminoethoxydiphenyl borate (2-APB), econazole and miconazole, soricidin and others) (Fecher-Trost et al., 2017; Haustrate et al., 2019).

Chapter 5

THERMO-TRP CHANNELS

The perception of temperature is a major component of sensory experience of animal and human organisms. Mammals evolved possessing protective mechanisms that facilitated survival in both cold and hot temperatures. A sensitive response of the nervous system to changes in temperature is of predominant importance for homeotherms to maintain a stable body core temperature. The central and periphery thermo-regulatory systems contain sensory receptors/signal transducers, integrators, and effector organs designed to regulate body temperature within a narrow range (Tipton et al., 2008; Taylor, 2014).

Several TRP channels exhibit highly temperature-dependent gating properties, which leads to steep changes in depolarizing current upon either cooling or heating. Based on this characteristic feature, these so-called 'thermo-TRPs' have been widely studied with the aim to elucidate their potential key role as thermo-sensors in the somatosensory system and to understand the basis of their high thermal sensitivity (Tominaga, 2009; Tsagareli, 2013, 2015; Voets, 2014).

5.1. TRPs in Central Thermosensation

The hypothalamic preoptic area (POA) is a region comprising the medial and lateral parts of the preoptic nucleus, the anterior hypothalamus, and nearby regions of the septum. It is recognized as a pivotal region in the central regulation of body temperature. Classic experiments have demonstrated that lesioning the POA abolished the ability of laboratory animals to mount proper thermoregulatory responses to changes in ambient temperature (see for review, Boulant, 2000). In addition, local heating or cooling of the POA induced the same physiologic thermoregulatory responses occurring when the temperature of the entire animal was increased or decreased, respectively (Conti, 2018).

A central role in POA thermoregulation is played by neurons known as warm-sensitive (WSN) that are identified by electrophysiology for their ability to increase firing rate with temperature above a certain thermal coefficient (Tabarean et al., 2005). It has recently provided the first evidence that at least some hypothalamic preoptic WSN have a functional marker that explains their ability to sense temperature by TRPM2 channel (Bartfai, 2016; Song et al., 2016).

Although TRP intrinsic thermo-sensitivity makes these channels in principle attractive as possible candidates for sensing temperature centrally and several have been reported to be expressed throughout the CNS, to date only TRPM2 has been demonstrated to have such a role. This was achieved by using calcium imaging based system to identify temperature-sensitive neurons within primary POA cultures in response to increased temperature that was up to 45°C for the initial screening. Very stringent criteria for the magnitude of response were adopted followed by educated selection and pharmacology to determine whether any TRP channel could participate in these responses in POA neurons. In addition to the channels known to respond to cold, TRPV1–3 and TRPM3 were discarded after their agonists failed to show any effect. By contrast, sensitization or activation of TRPM2 induced warm-sensitive currents (Conti, 2018).

The physiologic significance of TRPM2 in sensing central temperature became evident when comparing the fever response of mice lacking this receptor (*Trpm2*$^{-/-}$) with their wild-type counterpart (*Trpm2*$^{+/+}$) and following chemogenic regulation of *Trpm2* neurons. TRPM2, a channel that *in vivo* can respond to temperature ranges from 33 to 47°C depending on the cellular context, was activated in the POA at temperatures of 38°C or higher and is proposed to be instrumental in limiting excessively high fever. These findings suggest that activation of wild-type *Trpm2* neurons in a nonfever setting can induce hypothermia by increasing vasodilatation and inhibiting thermo-genesis in the brown adipose tissue. Of importance, the *Trpm2* neurons shown to have these properties were excitatory and acted, at least in part, by activating corticotrophin-releasing hormone neurons in the hypothalamic para-ventricular nucleus (Conti, 2018).

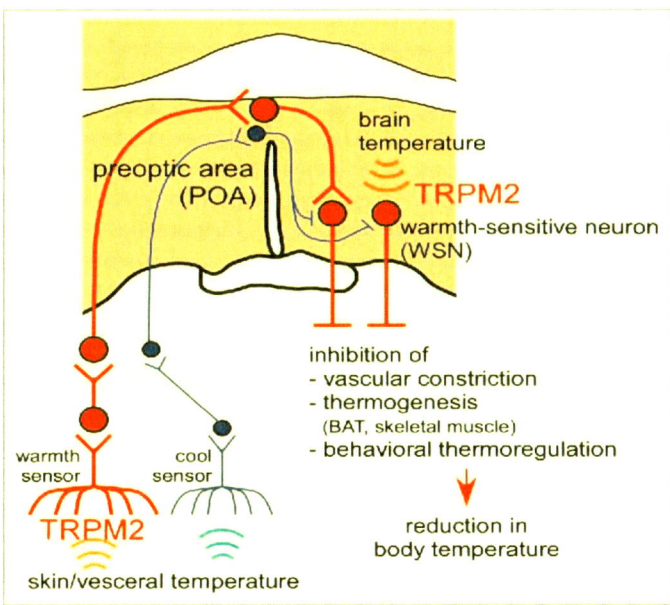

Figure 12. Schematic model for body temperature regulation involving TRPM2. Thermoregulation is driven by a neural network integrating thermal information *via* TRPM2 at skin/visceral warmth sensors and warmth-sensitive neurons (WSNs) in the preoptic area (POA). Warm temperatures detected by skin/visceral warmth sensors and WSNs promote heat dissipation by vascular dilation, inhibit thermogenesis in brown adipose tissue (BAT) and skeletal muscle and inhibit behavioral thermoregulation (seeking warmer place) thereby regulating whole body temperature (reproduced from Kashio, Tominaga, 2017, with permission).

As observed in WSNs of POA, heat-evoked responses in peripheral neurons were sensitized by H_2O_2-treatment, confirming the characteristics of TRPM2 as a thermo-sensor (Kashio et al., 2012). The temperature sensitivity of sensory neurons in the skin is responsible for detecting ambient temperature, mediating autonomic/behavioral thermoregulation (Morrison, Nakamura, 2011; Nakamura, Morrison, 2011). A thermal preference test showed that TRPM2 KO mice preferred warmer temperature in the innocuous temperature range (23 ~ 38°C) as compared with wild type (WT) mice, suggesting the roles of TRPM2 as a warmth sensor in skin to mediate behavioral thermoregulation (Figure 12) (Kashio, Tominaga, 2017; Señaris et al., 2018).

5.2. TRPs in Peripheral Thermal Sensitivity

Like other somatosensory modalities, innocuous peripheral warmth perception is mediated by a specialized subpopulation of cutaneous sensory neurons. Briefly, warmth-sensitive neurons exhibit low-level action potential firing at 'normal' skin temperatures (~ 29–32°C), but progressively increase their firing rates with increasing temperatures above this level. This response pattern makes warmth fibers more sensitive to elevated temperature than heat nociceptors, which mediate pain sensation, and tend to fire only at temperatures > 42–45°C. In addition, unlike heat nociceptors, which continue to encode temperatures well above 50°C, warmth-sensitive neurons typically, exhibit a plateau or even a reduction in firing rate at temperatures > 45°C. A third distinguishing characteristic of warmth-sensitive neurons is that, unlike some 'polymodal' heat-sensitive nociceptors, the former are invariably insensitive to mechanic stimulation of the skin (Jeon, Caterina, 2018).

To date, eleven thermo-sensitive TRP channels have been identified in mammals (Table 1).

These channels belong to the TRPV, TRPM, TRPA, and TRPC subfamilies, and their temperature thresholds for activation are in the range of physiological temperatures, which we can discriminate. TRPV1 and

TRPV2 are activated by elevated temperatures, whereas TRPM8 and TRPA1 are activated by cool and cold temperatures. TRPV3, TRPV4, TRPM2, TRPM4, and TRPM5 are activated by warm temperatures. TRPM3 was shown to be a sensor for noxious heat and TRPC5 was identified as a candidate cold sensor. Thermo-TRP channels expressed in sensory neurons and skin can act as ambient temperature sensors. Thermo-TRP channels usually function as 'multimodal receptors' that respond to various chemical and physical stimuli. Activation of these channels could contribute to changes in intracellular Ca^{2+} concentrations and control of membrane potentials in many cell types, except TRPM4 and TRPM5, which are not permeable of divalent cations. On the other hand, thermo-TRP channels are also expressed in tissues that are not exposed to dynamic temperature changes, suggesting that these channels have other physiological roles that are unrelated to sensation of temperature changes. Finally, recent studies have characterized the mechanism of thermo-sensing and the physiological role of thermo-TRPs in energy metabolism (Uchida et al., 2017).

Thermal sensing TRP channels can be divided into two subtypes, cold and heat receptors, based on response patterns. Cold sensitive thermo-TRP channels are characterized by an increase in the probability of opening with decreasing temperatures, while the heat receptors exhibit increased opening probability with increasing temperatures. On the other hand, response patterns of thermo-TRPs to thermal stimulations can be classified into four groups, i.e., cold, cool, warm, and heat, although no clear boundaries between cold and cool and between warm and heat have been defined experimentally (Liu et al., 2017).

As stated above, the TRP superfamily can be grouped into seven subfamilies based on structure and amino acid sequences with all TRP channels consisting of three structural domains, i.e., the N-terminal, transmembrane region, and the C-terminal (Figure 13). The transmembrane region contains six transmembrane segments (S1– S6) with varying degrees of sequence homology, and cation permeability (Venkatachalam, Montell, 2007). Both the N- and C- terminal domains are located intracellularly. The most distinctive segments are located on N-terminals, while the most conserved regions are found in the S6 domain, which is presumably the most important region for channel gating. Though many studies emphasize the importance of the N- and C-terminal domains of thermo-TRPs for the thermal senses, it is impossible to include these domains in the phylogeny-dependent selective analyses since these sequences are too diverse across subfamilies to fit the minimal demands of all models (Liu et al., 2017).

Table 1. Properties of thermo-TRP channels (reproduced from Uchida et al., 2017)

		Temperature threshold	Tissue distribution	Other Stimuli
Heat	TRPV1	> 42°C	Sensory neuron, brain, skin	Capsaicin, proton, capsiate, gingerol, shogaol, allicin, shanshool, camphor, resiniferatoxin, vanillotoxin, 2-APB, propofol, anandamide, arachidonic acid metabolic products (by lipoxygenases), monoacylglycerol, NO, extyracellular cation
	TRPV2	> 52°C	Sensory neuron, brain, spinal cord, lung, liver, spleen, colon, heart, immunocyte	Probenecid, 2-APB, cannabidiol, mechanical stimulation
Warm	TRPV3	> 32°C	Skin, sensory neuron, brain, spinal cord, stomach, colon	Camphor, carvacrol, menthol, eugenol, thymol, 2-APB
	TRPV4	> 27-41°C	Skin, sensory neuron, brain, kidney, lung, inner ear, bladder	4α-PDD, bisandrographolide, citric acid, arachidonic acid metabolic products (by epoxygenases), anandamide, hypoosmolality, mechanical stimulation
	TRPM2	> 36°C	Brain, immunocyte, pancreas etc.	(cyclic) ADPribose, β-NAD, H_2O_2, intracellular Ca^{2+}
	TRPM3	Warm-heat	Brain, sensory neuron, pancreas, eye	Ca^{2+} store depletion, pregnenolone sulfate, nifedipine, clotrimezole
	TRPM4	Warm	Heart, liver, immunocyte, pancreas etc.	Intracellular Ca^{2+}
	TRPM5	Warm	Taste cell, pancreas	Intracellular Ca^{2+}

		Temperature threshold	Tissue distribution	Other Stimuli
Cold	TRPM8	< 27°C	Sensory neuron	Menthol, icilin, eucalyptol
	TRPM5	Cold	Brain, sensory neuron, liver, heart, kidney	$G_{9/11}$-coupled receptors, diacylglycerol, GD^{3+}
	TRPA1	< 17°C	Sensory neuron, inner cell	Allyl isothiocyanate, carvacrol, cinnamaldehyde, allicin, dially trisulfide, miogadial, miogatrial, capsiate, acrolein, icilin tetrahydrocannabinol, menthol (10-100 µM), formalin, H_2O_2, alkalization, intracellular Ca^{2+}, NSAIDs, propofol/isoflurane/desflurane/etomidate/octanol/hexanol etc.

2-APB 2-aminorthoxydiphenyl borate, *NO* nitric oxide, *4α-PDD* 4α-phorbol-didecanoate, *ADPriibose* adenosine diphosphate ribose, *β-NAD* β-nicotinamide adenine dinucleotide, H_2O_2 hydrogen peroxide, *NSAIDs* non-steroidal anti-inflammatory drugs

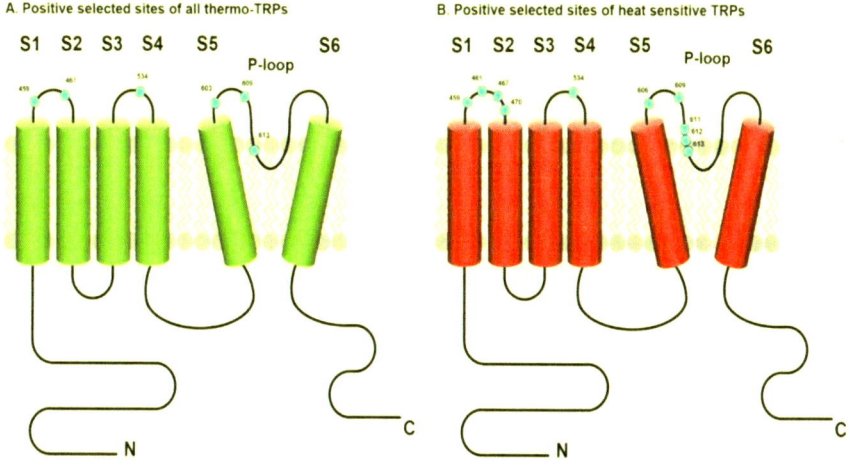

Figure 13. Positive selected sites of thermo and heat sensitive TRPs. Blue spheres represent positively selected sites which were numbered on the amino acid sequence of human TRPV1 (reproduced from Liu et al., 2017 with permission).

Based on this classification, it is common to speak of four thermal sensations (cold −10 to 15°C, cool 16–30°C, warm 31–42°C and hot 43–60°C), whereby cold and hot are potentially noxious and/or painful (Figure 14).

5.2.1. TREK Channels

The TWIK-related potassium channel (TREK) subfamily belongs to the two-pore domain potassium channels family (K2P) and is comprised of three members: TREK1, TREK2, and TRAAK (TWIK-related arachidonic acid-activated potassium channel). These are background K^+ channels characteristically modulated by several physical and chemical stimuli, such as membrane stretch, pH, unsaturated fatty acids, general anesthetics, and temperature. In general, TREK channels display very weak activity at room temperature and normal pressure, even when over-expressed in heterologous systems. However, their activity increases strongly when a number of different stimuli are applied, including an increase in temperature. From a physiological point of view, it is important to note that at 37°C, all three members of the TREK subfamily respond to stimuli (pH, membrane stretch, or arachidonic acid), much like they do at room temperature. TREK channels may fulfill a dual role in the transmission of thermal pain. Thus, their strong activation by noxious heat results in an outward current that provokes membrane hyperpolarization and a reduction of thermo-receptor firing, provoking heat-pain relief. Conversely, inhibition of TREK channels by noxious cold should depolarize thermo-receptors and increase their excitability, cooperating in the transduction of noxious cold sensations (Lamas et al., 2019).

5.2.2. Heat-Sensitive TRP Channels

Four TRP subtypes are activated by an increase in temperature (Figure 14). Two of them respond to warm stimuli (TRPV4, warm > 27°C, and TRPV3, warm > 34°C), and the other two to hot-painful stimuli (TRPV1, hot > 43°C, and TRPV2, hot > 52°C).

TRPV1s are voltage- and temperature-dependent channels that display outward rectification when expressed in human embryonic kidney (HEK) cells and that are strongly enhanced by heating (to 48°C) and by capsaicin. At room temperature, the current passing through these channels is negligible below 0 mV, but at 42°C the channel activates more or less between − 100 and +50 mV. These cationic channels are ten times more permeable to Ca^{2+} than to Na^+ (PCa/PNa ~10) and are thought to be sensors for noxious heat but not activated by innocuous heat. It is interesting to note that inflammatory mediators like ATP and bradykinin strongly reduce the threshold of TRPV1 activation (30°C) such that warm temperatures become painful. TRPV1 is strongly expressed in small-diameter sensory neurons of the DRG, trigeminal ganglion (TG), and nodose ganglion (NG), but also in the hypothalamus, sites where they may exert an important role in thermo-reception (Jeon, Caterina, 2018; Lamas et al., 2019).

TRPV2 is activated at extremely high temperatures (52°C), although it is not affected by capsaicin and shows a PCa/PNa ~3. It is thought that the temperatures that activate TRPV2 are more harmful than those that activate TRPV1. These channels are strongly expressed by myelinated medium-large diameter DRG neurons ($A\delta$ and $A\beta$), as well as in the hypothalamus and in the NG (Nedungadi et al., 2012).

TRPV3 channels are activated at warm, close to hot, temperatures (around 34–39°C), generating currents with pronounced outward rectification and a PCa/PNa ~ 12. They are capsaicin-insensitive channels but stimulated by camphor, and they are thought to be involved in thermo-sensation and thermal nociception. Indeed, it has been suggested that TRPV3 channels contribute more to the speed with which mice select a more comfortable temperature than to the choice of the value of the temperature itself. Interestingly, it was proposed that TRPV3 channels transmit thermal stimuli through skin keratinocytes, which in turn will transmit this information to sensory endings. TRPV3 channels are expressed in sensory DRG and NG neurons but also in the hypothalamus and it co-localizes with TRPV1 in DRG neurons (Lamas et al., 2019).

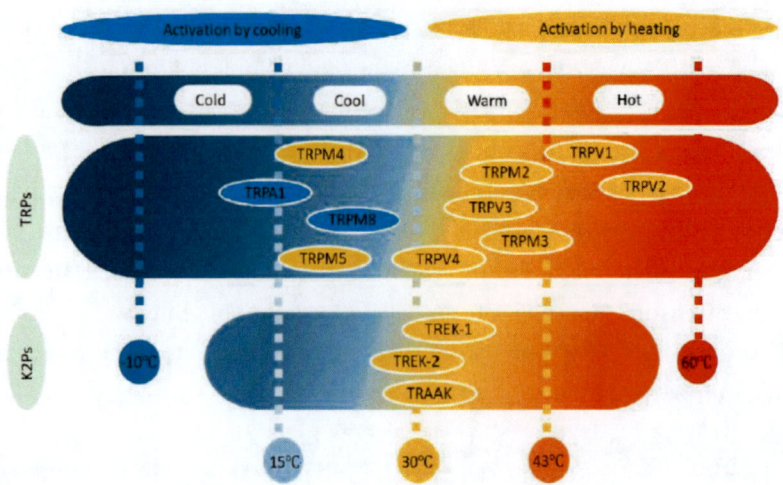

Figure 14. Distribution TRP and TWIK-related potassium (TREK) channels as a function of their temperature threshold. Note that while TREK channels are activated by increases in temperature (orange), TRP channels may also be activated by lowering the temperature (blue) (reproduced from Lamas et al., 2019).

TRPV4 are cationic (PCa/PNa ~ 6) channels activated at even lower warm temperatures (around 27°C), generating outwardly rectifying currents and responding dynamically to temperature changes in the physiological range. These channels were proposed to play a role in thermosensation and thermo-regulation (Todaka et al., 2004). Similarly, some behavioral studies reported a reduced response to temperature changes in TRPV4 KO mice. TRPV4 channels are expressed in DRG, TG, NG, and preoptic/anterior hypothalamic neurons (Lamas et al., 2019) although in the hypothalamus they seem to be expressed in terminals rather than in the soma, such that their role in body thermoregulation is unclear (Wechselberger et al., 2006).

TRPM2 (> 35°C), TRPM3 (> 40°C), TRPM4 (> 15°C), and TRPM5 (> 15°C) are channels that can also be activated by warming (Figure 14), yet they have received less attention, probably because it was initially thought that they were not expressed by somatosensory neurons or keratinocytes (Talavera et al., 2005). TRPM2 as a voltage-insensitive, shows a PCa/PNa ~1, and activates at 35°C, (Tan, McNaughton, 2016).

TRPM3 is expressed broadly, generating an outwardly rectifying current, having a PCa/PNa between 0.1 and 10, and activating at > 40°C (Thiel et al., 2017; Vriens et al., 2011). It is important to say that TRPM3 has been described as part of a triad of TRPs, together with TRPV1 and TRPA1, involved in the transduction of acute noxious heat in mice (Vandewauw et al., 2018). The combined ablation of these channels (triple KO) was necessary for the complete reduction of acute noxious sensing; single or double KO combinations resulted in deficits in heat responsiveness, but mice still conserved vigorous withdrawal responses to noxious heat (Vandewauw et al., 2018).

The mechanism by which temperature modulates TRP channels is still unclear, yet several hypotheses have been proposed: (1) changes in temperature could produce a ligand that binds to a receptor and affects the channel; (2) changes in temperature could produce a structural change in the channel that provokes its opening; (3) temperature changes could affect the structure of the membrane, causing changes in tension that would in turn affect ion channels. Because capsaicin induces burning pain, it has been hypothesized that both capsaicin and heat may use a common mechanism to activate TRPV1 and produce pain. Both stimuli affect excised patches, and in general, it is accepted that TRPV1 is directly activated by noxious-heat, so that it can be considered a true heat sensor (Lamas et al., 2019).

5.2.3. Cold-Sensitive TRP Channels

Two TRP channels are activated by decreases in temperature (Figure 14), TRPM8 (< 25°C) activates in the cool range while TRPA1 (< 18°C) senses cold-painful temperatures. Similarly, cool fibers (Aδ and C) have activation thresholds at about 30°C and cold fibers (C) have activation thresholds < 20°C. Accordingly, two populations of TG neurons were described in terms of their activation threshold when temperatures decrease at 30 and 20°C for a low and high threshold, respectively. In general, cold fibers fire continuously at normal skin temperatures and they increase their firing frequency when the skin is cooled down, or they shut down when the skin is warmed. In addition, cold fibers can adapt to small decreases in temperature (Lamas et al., 2019).

In a recent study, the cellular basis for noxious cold detection was identified, implicating the TRP genes *Pkd2*, *NompC*, and *Trpm* in regulating cold-evoked nociceptive behavioral responses. When challenged with temperatures at and below 14°C, *Drosophila* larvae execute a full-body contraction behavior along the head-to-tail axis. Optogenetic activation and synaptic blockade analyses reveal class III neurons of *D. melanogaster* function as primary noxious cold nociceptors required for cold-evoked contraction response. Moreover, *in vivo* Ca^{2+} imaging demonstrates class III neurons are specifically activated by cold stimulation (Himmel, Cox, 2017; Himmel et al., 2017; Turner et al., 2016).

TRPM8 channels are voltage-dependent cationic channels that are permeable to Na^+, K^+, Cs^-, and Ca^{2+} (PCa/PNa ~3). When expressed in HEK cells and recorded in whole-cell configuration, they show a voltage-dependent outwardly rectifying current that strongly increases upon cooling from 30 to 15°C or through the application of menthol. Importantly, both basal and cold-stimulated currents reverse around 0 mV and were almost negligible below this potential. Cooling CHO cells expressing TRPM8 (in the range of 25 to 15°C) also induces an increase in intracellular calcium. The effect of temperature is due to an increase in the open probability and a shift in the conductance–voltage relationships along the voltage axis. Similar results were obtained in inside-out macro-patch recordings, although the stimulation occurred at lower temperatures, suggesting that the integrity of the cell is important but not indispensable (Voets et al., 2004). The role of this channel as a detector of painful cold has been questioned in experiments on KO mice, but nevertheless, it is accepted that it is an important cold sensor in vagal, TG, and especially DRG afferents (Lolignier et al., 2016; Señarís et al., 2018). It was predicted that cold transduction may require the activation and inhibition of several different ion channels, such as TRP, TREK, and ENaC channels. If this were the case, TRP channels would probably be more important in the noxious-cold range, whereas TREK channels might participate more strongly in the cool range of temperatures. TRPM8 is expressed in small-diameter DRG and TG neurons, presumably thermo-receptors, yet it seems not to co-localize with TRPV1 (Lamas et al., 2019).

TRPA1 is activated by lower temperatures than TRPM8 (< 18°C), and while it would be expected to be involved in cold nociception, this is not that clear. TRPA1 generates an outward rectifying cationic current, both in control conditions and when cold activated (about 10°C), with similar permeability for Ca^{2+} and Na^+ (PCa/PNa ~ 1). Cinnamaldehyde can selectively activate currents through this channel in native DRG neurons, as can bradykinin (when TRPA1 co-expressed with bradykinin receptors), strongly suggesting a role in sensing nociceptive stimuli. However, TRPA1 KO mice do not seem to have difficulties in sensing cold stimuli through the skin, while the response of TRPM8 KO mice to cold is significantly dampened. By contrast, about 50% of NG neurons in culture were activated by cooling (< 24°C), mainly through TRPA1 channel activation (increase in $[Ca]_i$, depolarization, and action potential firing). Interestingly, about 10% of the NG neurons responded to cold through a TRPA1- and calcium independent pathway. TRPA1 often co-localizes with TRPV1, and in fact, this could explain the paradoxical hot sensation experienced with an extremely cold stimulus. Interestingly, most NG neurons sensitive to cold are also sensitive to heat. TRPV1 is expressed in DRG and TG, while TRPA1 and TRPM8 are not co-expressed in DRG neurons, but they are in TG neurons. The data available, thus, suggest that TRPA1 is the principal ion channel involved in cold sensation in visceral NG neurons, while TRPM8 would fulfill the same role in somatic neurons (Lamas et al., 2019).

In the recent review Viana (2016) highlights the role of TRPA1 as a multipurpose sensor of harmful signals, including toxic bacterial products and ultra violet light, and as a sensor of stress and tissue damage. Sensing roles span beyond the peripheral nervous system to include major barrier tissues: gut, skin and lung. Tissue injury, environmental irritants and microbial pathogens are danger signals that can threaten the health of organisms. These signals lead to the coordinated activation of the nociceptive and the innate immune system to provide a homeostatic response trying to re-establish physiological conditions including tissue repair. Activation of TRPA1 participates in protective neuro-immune interactions at multiple levels, sensing reactive oxygen species (ROS) and bacterial products and triggering the release of neuropeptides (Viana, 2016).

The fact that TRPM8 can be activated by cooling in inside-out patches suggests that the mechanism is membrane delimited, also arguing against the participation of a second messenger pathway. Notwithstanding, inhibition of phospholipase C strongly dampened the increase in calcium provoked by cold stimuli in TRPM8-expressing CHO cells. Cooling activates TRPM8-expressed channels by causing a shift in the voltage dependence of activation to negative values, and the same mechanism is responsible for the activation of TRPV1 by heat but not the activation of TRPM2 (Lamas et al., 2019). It has been proposed that temperature induces large rearrangements of the protein and thus, the existence of a temperature-sensing domain or 'temperature sensor' in the structure of TRPM8 channels. Much like for TRPM8, inhibition of phospholipase C strongly reduces the increase in calcium provoked by cold stimuli in TRPA1-expressing CHO cells (Vandewauw et al., 2018).

Chapter 6

TRP CHANNELS IN PHYSIOLOGICAL NOCICEPTION AND PAIN

The nervous system detects and interprets a wide range of thermal and mechanical stimuli, as well as environmental and endogenous chemical irritants. When intense, these stimuli generate acute pain, and in the setting of persistent injury, both peripheral and central nervous system components of the pain transmission pathway exhibit tremendous plasticity, enhancing pain signals and producing hyper-sensitivity. When plasticity facilitates protective reflexes, it can be beneficial, but when the changes persist, a chronic pain condition may result. Recent genetic, electrophysiological and pharmacological studies are elucidated the molecular mechanisms that underlie detection, coding, and modulation of noxious stimuli that generate pain (Basbaum et al., 2009).

As we stated above the TRP channel superfamily is comprised of a large group of cation-permeable channels, which display an extraordinary diversity of roles in sensory signaling and are involved in plethora of animal behaviors. These channels are activated through a wide variety of mechanisms and participate in virtually every sensory modality. Although the physiological functions of most TRP channels are not well known, their wide distribution indicates that the biological roles and activation mechanisms of these channels are very important for living bodies. For instance, TRP channels are best recognized for their contributions to sensory transduction, response to temperature, nociceptive and itching stimuli, touch, osmolarity, chemical substances such as pheromones or odorants, and other stimuli from both within and outside the cell.

Analysis of TRP channel function and expression has validated the existence of nociceptors as a specialized group of somatosensory neurons devoted to the detection of noxious stimuli. These studies are also providing insight into the coding logic of nociception and how specification of nociceptor subtypes underlies behavioral discrimination of noxious thermal, chemical, and mechanical stimuli (Julius, 2013). It is interesting that several TRP channels exhibit sensitivity to substances previously known to cause pain or pain-like sensations; these include pungent natural compounds present in cinnamon oil, wintergreen oil, clove oil, mustard oil, ginger oil, chili pepper, mint and which are menthol, capsaicin, camphor, thymol, and others. This data clearly indicates that the specific and selective inhibition of TRP channels can be used to relieve pain and thus these channels represent promising targets for the development of further generation of analgesic drugs (Benemei, Dussor, 2019; Dietrich, 2019; Geppetti et al., 2015; Hantute-Ghesquier et al., 2018; Samanta et al., 2018; Vangeel, Voets, 2019).

Almost fifteen years ago in collaboration with Earl Carstens laboratory from University of California at Davis, we started a study of agonists-evoked activation of thermo-TRP channels using a battery of behavioral tests. Firstly, we carried out experiments on thermal hyperalgesia and mechanical allodynia induced by intradermal cinnamaldehyde and menthol (Carstens et al., 2010; Klein et al., 2010; Tsagareli et al., 2010) and later by allyl isothiocyanate (a main compound of mustard oil) in rats (Nozadze et al., 2016a,b, 2019).

6.1. TRPA1 Channel

A number of temperature-sensitive TRP ion channels have been studied as nociceptors that respond to extreme temperatures and harmful chemicals. Strong activation of these channels in the nervous system by their agonists (cinnamaldehyde, mustard oil, farnesyl thiosalicylic acid, acrolein, 4-hydroxynonenal, formalin, and others) elicits pain. One member of the TRP family, the ankyrin-repeat channel TRPA1 is of special interest for its functional diversity as a sensor of irritating and cell damaging signals, and its important role in pain and itch and of many different diseases (Nilius et al., 2012; Nozadze et al., 2019; Tsagareli et al., 2019; Zheng, 2013).

6.1.1. Cinnamaldehyde-Induced Activation of TRPA1 Channel

Cinnamaldehyde (CA) is a pungent chemical from cinnamon that acts as an agonist of the thermo-TRP ion channel TRPA1 that was originally reported to be activated by warm temperatures below 18°C (Story et al., 2003). TRPA1 is activated by several other irritant chemicals including mustard oil (allyl isothiocyanate, AITC) (Bautista et al., 2006; Jordt et al., 2004; McNamara et al., 2007) and is co-localized with TRPV1 in polymodal nociceptors (Kobayashi et al., 2005; Story et al., 2003). Topical application of CA or AITC elicits a burning sensation and heat hyperalgesia in humans (Albin et al., 2008; Namer et al., 2005; Simons et al., 2003) and enhances responses of rat spinal dorsal horn and trigeminal subnucleus caudalis (Vc) neurons to noxious heat (Merrill et al., 2008; Simons et al., 2004; Zanotto et al., 2008). The role of TRPA1 in cold pain is more controversial, with discrepant reports that TRPA1 does (Karashima et al., 2009; Story et al., 2003) or does not respond to intense cooling (Jordt et al., 2004). Knockout mice lacking TRPA1 exhibit normal cold sensitivity (Bautista et al., 2006), or partial (Kwan et al., 2006) or severe (Karashima et al., 2009) deficits in cold pain sensitivity. In humans, topical cutaneous application of CA induces cold hypo-algesia (Namer et al., 2005) while epilingual CA induces brief cold hyper-algesia (Albin et al., 2008). Neither agent affects rat spinal neuronal responses to cooling (Merrill et al., 2008; Sawyer et al., 2009).

TRPA1 was originally reported also to play a role in mechano-transduction (Corey et al., 2004) and AITC induced mechanical allodynia in humans (Koltzenburg et al., 1992). Intraplantar injection of the TRPA1 agonist 4-hydroxynonenal reduced mechanical paw withdrawal thresholds in mice (Trevisani et al., 2007) and blockade of TRPA1 by systemically or locally administered antagonists reversed mechanical hyperalgesia in inflammatory and nerve injury models in mice (Eid et al., 2008; Petrus et al., 2007).

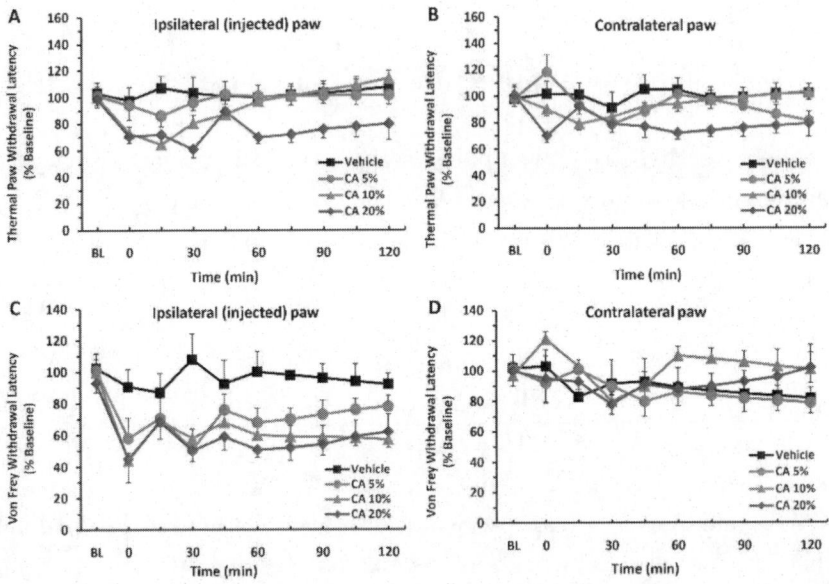

Figure 15. CA-induced heat hyperalgesia and mechanical allodynia. (A): Normalized mean thermal paw withdrawal latency *vs.* time following ipsilateral intraplantar injection of vehicle (control) or CA at each indicated concentration. There was a significant effect of CA concentration (repeated measures ANOVA, $F_{3,28} = 3.1$, $p < 0.05$) with the 20% CA group significantly different from vehicle and 5% CA groups (LSD; $p < 0.05$ for both). BL: pre-injection baseline. Error bars: SEM (n = 8). (B): As in (A) for paw contralateral to CA injection. There was a significant effect of CA concentration ($F_{3,28} = 3.13$, $p < 0.05$) with the 20% group significantly different from vehicle and 5% groups (LSD; $p < 0.05$ for both). (C): As in (A) for von Frey mechanically-evoked withdrawal of the injected paw. There was a significant effect of CA concentration ($F_{3,28} = 4.4$, $p < 0.05$) with 5%, 10% and 20% CA all different from vehicle but not from each other. (D): As in (C) for contralateral paw. There was no significant effect of CA concentration ($F_{3,28} = 0.14$, $p > 0.5$), (Reproduced from Tsagareli et al., 2010).

Our obtained data showed that CA resulted in a significant, dose-dependent reduction in ipsilateral thermal paw withdrawal latency. Figure 15(A) shows mean withdrawal latencies of the injected paw *vs.* time relative to injection of vehicle or CA at each concentration tested. There was a dose-dependent reduction in latency, with the 20% CA concentration significantly different from vehicle and 5% CA treatments. The highest dose resulted in a mean reduction to 61.7% of pre-injection baseline by 30 min with partial recovery at 120 min. For the contralateral paw (Figure 15B), there was an overall significant effect of treatment with the 20% group being significantly different from saline and 5% CA treatments. Mechanically evoked withdrawal thresholds are plotted *vs.* time for the treated paw in Figure 15(C). At each CA concentration thresholds were significantly different from vehicle, but not from each other, indicating that a maximal reduction in withdrawal threshold (to 44.4% of baseline) was achieved at the lowest (5%) concentration of CA. Mean withdrawal thresholds for the contralateral paw were not significantly affected at any CA concentration (Figure 15D).

In the 30°C *vs.* 15°C thermal preference test, rats treated with the highest CA concentration exhibited a small but significant avoidance of the colder plate, spending a significantly ($p < 0.001$, paired *t*-test) greater percentage of time on the warmer 30°C plate ($83.3 \pm 6.9\%$ SEM). Almost the same difference is for vehicle-treated rats compared to the 30°C ($76.5 \pm 7.1\%$) plate *vs.* 15°C plate (Figure 16).

When tested on the −5°C cold plate test (3 min post-treatment), there was no significant difference in latency between groups treated with the highest concentration of CA compared to vehicle (26.0 ± 5.3 s *vs.* 24.9 ± 2.24 s; $p > 0.05$, $n = 8$) (data not shown). In the 0°C cold plate test, CA treatment resulted in a significant concentration- and time-dependent decrease in latency indicative of cold hyperalgesia (Figure 17A). The mean latency was significantly different between 5% CA and 10% or 20% CA treatment groups at the initial (3-min) time point ($p < 0.05$, repeated measures ANOVA). When tested on the +5°C cold plate, many animals did not display nocifensive behavior up to the 150 s cutoff time consistent with a previous report. CA treatment resulted in a significant reduction in latency (Figure 17B) with no significant effect of CA concentration over time, indicating that a maximal effect was achieved at the lowest CA concentration.

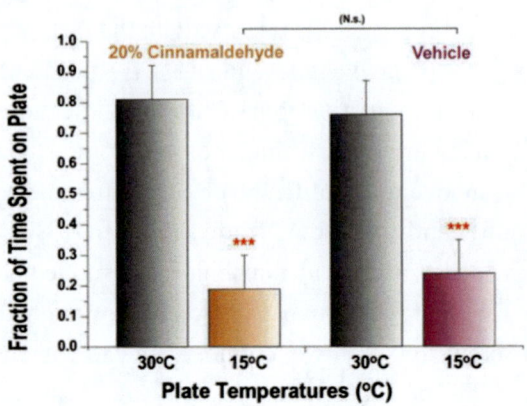

Figure 16. Experimental and control groups of rats injected with CA show a significant avoidance of the colder plate.

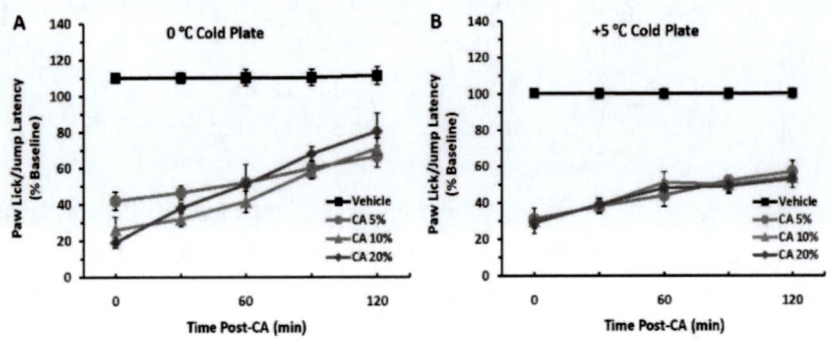

Figure 17. Intraplantar CA decreases cold plate latency. Graphs plot normalized mean latencies (error bars: SEM; n = 8) to lick hindpaw or jump after being placed on a 0 °C (A) or +5°C (B) cold plate surface over a 2 h period of time. (A): The mean latency was significantly different between 5% CA *vs.* 10% or 20% CA treatment groups at the first (3-min) time point (repeated measures ANOVA, $p < 0.05$). (B): There was a significant reduction in latency for each CA treatment group with no significant effect of CA concentration over time (reproduced from Tsagareli et al., 2010).

The presented data provide a comprehensive view of effects of intraplantar CA on thermal (hot and cold) and mechanical sensitivity. CA induced a dose-dependent heat hyperalgesia lasting > 2 h at the highest dose, mechanical allodynia, and cold hyperalgesia. CA enhancement of heat sensitivity is consistent with previous studies. Topical application of CA (795mM) to human forearm skin evoked burning pain and heat hyperalgesia (Namer et al., 2005). Epilingual CA (16mM) produced brief heat hyperalgesia (Albin et al., 2008). CA enhanced responses of spinal (Merrill et al., 2008) and Vc neurons to noxious heat (Zanotto et al., 2008). These and the present findings are consistent with a role for TRPA1 in heat pain and heat hyperalgesia.

The dose-dependent increase in magnitude and duration of heat hyperalgesia induced by CA was similar to that induced by intraplantar capsaicin (1–30 µg dose range) in rats using the same method (Gilchrist et al., 1996). Since TRPA1 is co-expressed in sensory neurons that express TRPV1 (Kobayashi et al., 2005; Story et al., 2003), heat hyperalgesia induced by CA might involve its activation of intradermal nociceptor nerve endings to engage an intracellular mechanism leading to enhanced heat sensitivity of TRPV1. Alternatively, CA may cause intradermal release of inflammatory mediators that lower the heat threshold of TRPV1 (Chuang et al., 2001; Sugiura et al., 2002). CA at the highest concentration may have also triggered central sensitization, leading to the observed reduction in withdrawal latency for the contralateral paw (Figure 15B). Consistent with this, topical application of AITC (TRPA1 agonist) to the lateral hindlimb significantly reduced the tail flick latency in rats in a manner dependent on the integrity of the rostral ventro-medial medulla (Heinricher, Fields, 2013; Ren, Dubner, 2009; Urban et al., 1999).

CA significantly lowered nocifensive response latencies in the 0°C and +5°C cold plate tests, which we interpret as cold hyperalgesia (Figure 17). The lack of effect of CA in the −5°C cold plate test may reflect a floor effect, since the response latency at this temperature was maximal and thus could not be reduced further by CA. CA also weakly but significantly increased cold avoidance in the thermal preference test. The latter may reflect cold allodynia, on the assumption that the 15°C surface is normally innocuous but became painful following CA treatment. Cold avoidance progressively increases as the temperature decreases from 25°C to 0°C, and it is possible that following CA treatment the 15°C surface was perceived to be colder and hence aversive but not necessarily painful.

In humans, CA on forearm skin induced cold hypo-algesia (Namer et al., 2005), whereas epilingual CA or AITC briefly enhanced cold pain (Albin et al., 2008). Lingual application of CA significantly enhanced cold-evoked responses of superficial Vc neurons in rats (Zanotto et al., 2008) but did not affect responses of lumbar spinal wide dynamic range (WDR)-type neurons to skin cooling (Merrill et al., 2008). These discrepancies regarding the effects of CA on cold pain perception and neuronal responses may partly involve the route of delivery and accessibility of CA to intradermal nociceptors. In the present study, intradermal injection of CA allowed for a direct access to nociceptive nerve endings to result in significant cold hyperalgesia and enhancement of cold avoidance.

The prolonged enhancement of mechano-sensitivity (i.e., allodynia) following CA (Figure 15C) is consistent with previous studies showing a prolonged decrease in mechanical withdrawal threshold in mice following intraplantar injection of 4-Hydroxynonenal, a TRPA1 agonist (Trevisani et al., 2007), and with allodynia induced in human skin by topical application of AITC (Koltzenburg et al., 1992). A role for TRPA1 in mechanical allodynia is further supported by reports that TRPA1 antagonists attenuated inflammation- or nerve injury-induced decreases in mechanical paw withdrawal thresholds in mice (Eid et al., 2008; Petrus et al., 2007) and decrease mechanically evoked responses in C fibers in mice (Kerstein et al., 2009). However, these behavioral data are inconsistent with our electrophysiological data showing that neither CA nor AITC had any significant effect on mechanical sensitivity of spinal WDR neurons (Merrill et al., 2008). Similarly, only 1 of 9 subjects experienced mechanical allodynia following application of 10% CA to forearm skin (Namer et al., 2005). The mismatch between our behavioral observation of a CA-induced increase in mechano-sensitivity and lack of CA effect on neuronal mechano-sensitivity (Merrill et al., 2008) may involve the route of administration as noted earlier.

In conclusion, these data are consistent with roles for TRPA1 in thermal (hot and cold) hyperalgesia and mechanical allodynia (Carstens et al., 2010; Tsagareli et al., 2010).

6.1.2. Mustard Oil-Induced Activation of TRPA1 Channel

Allyl isothiocyanate (AITC) is the organo-sulfur compound that is responsible for the pungent taste of mustard, horseradish, and wasabi. This pungency and the lachrymatory effects of AITC are mediated through the thermo-sensitive TRPA1 and TRPV1 ion channels. AITC from mustard and allicin from garlic are the archetypal dietary TRPA1 activators (Laursen et al., 2015; Nilius, Appendino, 2013; Nilius et al., 2012).

In our experiments injections of AITC in rats' hindpaw resulted in a significant dose dependent reduction in the ipsilateral thermal paw withdrawal latency. Figure 18A shows the mean withdrawal latencies of the injected paw *vs.* time relative to injection of vehicle or AITC at each concentration tested. There was a dose-dependent reduction in the latency, with the 15% AITC concentration being significantly different from vehicle and 5% AITC treatments. The highest dose resulted in a mean reduction to 73.7% of the pre-injection baseline value by 30 min. For the contralateral paw, there was an overall significant effect of treatment, with the 15% group being significantly different from the vehicle group (Figure 18B).

Mechanically evoked withdrawal thresholds were plotted *vs.* time for the treated paw in Figure 18C. At each AITC concentration, with partial recovery at 120 min, the thresholds were significantly different from those with vehicle, but not from each other. The mean withdrawal thresholds for the contralateral paw were not affected significantly at any AITC concentration (Figure 18D).

Two-temperature preference test showed that on 30°C *vs.* 15°C plates, rats treated with higher (10% and 15%) concentrations of AITC exhibited a significant preference for the colder (15°C) plate ($P < 0.05$ and $P < 0.01$, respectively) compared with vehicle-treated and 5% concentration-treated rats (Figure 19) – that is, the animals significantly avoided the warmer (30°C) plate.

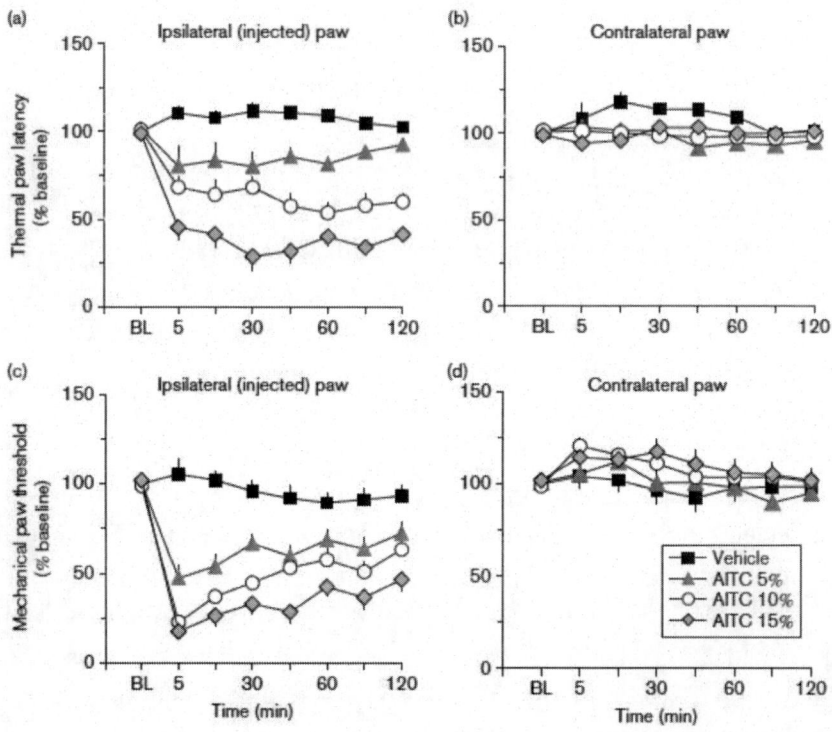

Figure 18. (A) Thermal paw withdrawal latency versus time following ipsilateral intraplantar injection of vehicle (control) and AITC at each of the indicated concentrations. There was a significant effect of all concentrations of AITC compared with vehicle ($P < 0.001$). (B) As in (A) for the paw contralateral to AITC injection. There were no significant effects of any AITC concentration compared with vehicle. (C) As in (A) for Von Frey mechanically evoked withdrawal of the injected paw. There were significant effects of all AITC concentrations relative to vehicle ($P < 0.001$). (D) As in (C) for the contralateral paw, without a significant effect of any AITC concentration. N = 6/group. AITC, allyl isothiocyanate; BL, pre-injection baseline (adapted from Nozadze et al., 2016b).

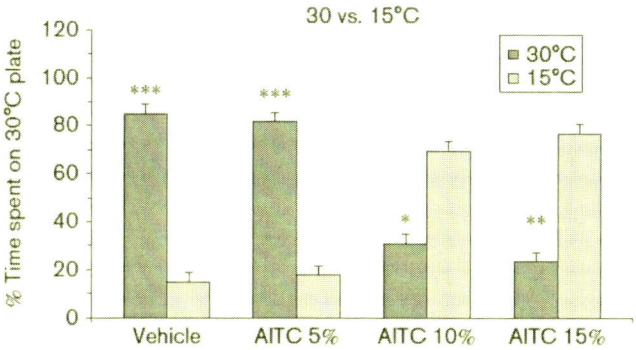

Figure 19. Biphasic effects of AITC on thermal preference. The figure shows the % of time for which rats stood on 30 vs.15°C plates; n = 6/group. At high (10 and 15%) AITC concentrations, the rats spent significantly more time on the 15°C plate (light gray bars) compared with the 30°C plate (dark gray bars; P < 0.05 and < 0.01, respectively), indicating cold hypo-sensitivity. At the lower (5% and vehicle) concentration, rats spent significantly more time on the 30°C plate (P < 0.001), indicating cold hyper-sensitivity. $^*P < 0.05$; $^{**}P < 0.01$; $^{***}P < 0.001$ (adapted from Nozadze et al., 2016b).

Bilateral intraplantar injection of AITC induced a significant reduction in cold plate latency compared with the vehicle control, which we interpret as a cold hyperalgesia. However, there were no antinociceptive effects on latency at the +5, 0 and −5°C temperatures. In the −5°C cold plate test, AITC treatment resulted in a highly significant difference between the vehicle-treated (mineral oil) and AITC-treated groups (P < 0.001) (Figure 20). Note that a vehicle solution shows some protective effects against cold temperatures, especially against −5°C in the mineral oil control group, compared with AITC injections (Figure 20C).

The presented data provide a detailed view of the effects of intraplantar injections of AITC on thermal and mechanical sensitivities. These influences induced heat hyperalgesia and mechanical allodynia lasting more than 2 h. The AITC-induced enhancement of heat sensitivity is consistent with the findings of our previous studies with CA injections in rats (Carstens et al., 2010; Tsagareli et al., 2010; Tsagareli, 2011) and on applications of this agent in human participants. In particular, AITC, and

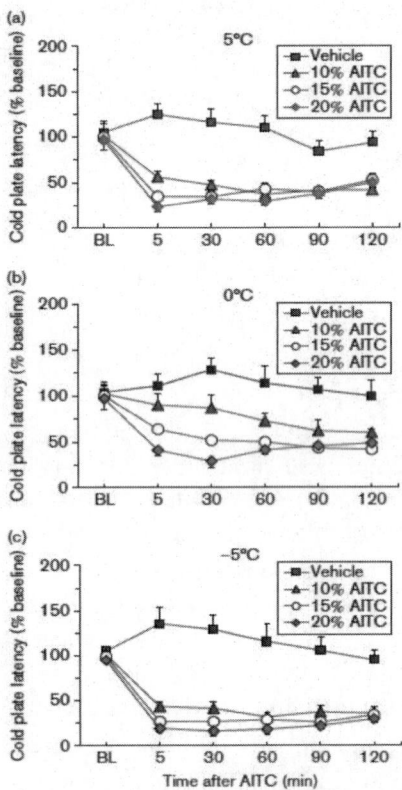

Figure 20. Intraplantar AITC injections produce dose-dependent cold hyperalgesia at temperatures of (a) +5°C, (b) 0°C and (c) −5°C, respectively. The figure shows a change in cold plate latency (% of pre-AITC baseline) *vs.* time after topical AITC application at the concentrations indicated; n = 12/group. Note that the effect of AITC is maximal at the first time point (5 min), as AITC diffuses readily through the skin (adapted from Nozadze et al., 2016b).

CA enhanced lingual heat pain elicited by a 49°C stimulus. At the same time AITC and CA weakly enhanced lingual cold pain (9.5°C), whereas capsaicin had no effect (Albin et al., 2008). Other investigators found that topical application of AITC produces neurogenic inflammation and, concurrently, heat and mechanical hyperalgesia, presumably through a centrally mediated sensitization process, and that these effects are TRPA1-mediated (Bandell et al., 2004; Macpherson et al., 2005; Bautista et al., 2006; Kwan et al., 2006; Belmonte, Viana, 2008; Bráz, Basbaum, 2010).

Dose-dependent increases in the magnitude and duration of heat hyperalgesia induced by AITC and CA were similar to those induced by intraplantar capsaicin. As TRPA1 is co-expressed in sensory neurons expressing TRPV1 (Kobayashi et al., 2005), heat hyperalgesia induced by AITC and CA might involve activation of these receptors (sensory intradermal terminals of nociceptor nerve endings) through an intracellular mechanism, leading to enhanced heat sensitivity of TRPV1.

The finding of long-lasting enhancement of mechano-sensitivity (i.e., allodynia) following CA and AITC applications (Figures 15C and 18C) is consistent with previous studies that showed a prolonged decrease in the threshold for mechanical withdrawal in mice following intraplantar injection of a TRPA1 agonist, bradykinin (Chuang et al., 2001; Sugiura et al., 2002), and with studies showing the induction of allodynia in human skin by topical application of AITC (Albin et al., 2008). The role of TRPA1 in mechanical allodynia is further supported by reports that a TRPA1 antagonist, 4-hydroxynonenal, attenuated inflammation-induced or nerve injury induced decreases in mechanical paw withdrawal thresholds and decreased mechanically evoked responses in C-fibers in mice (Trevisani et al., 2007). In addition, AITC and CA were shown to induce mechanical allodynia in humans (Koltzenburg et al., 1992). However, TRPA1, as a ligand-gated ion channel in sensory neurons, was initially reported to be activated by cold temperatures, below 18°C (Namer et al., 2005; Albin et al., 2008; Caterina, 2007), although this opinion has been disputed (Jordt et al., 2004; Belmonte, Viana, 2008). TRPA1-knockout mice exhibited either normal cold sensitivity (Belmonte, Viana, 2008) or mild (Albin et al., 2008) or severe deficits (Namer et al., 2005) in cold sensitivity in human participants.

It has been recently shown that TRPA1 channels in the skin contribute to sustained and noxious mechanical stimulus evoked postoperative pain, whereas spinal TRPA1 channels contribute predominantly to innocuous mechanical stimulus-evoked postoperative pain (Pertovaara, Koivisto, 2011; Wei et al., 2012). Furthermore, spinal TRPA1 receptors are responsible for central pain hypersensitivity under various pathophysiological conditions, such as inflammatory and neuropathic pain

(Eid et al., 2008; Kerstein et al., 2009; Wei et al., 2011). However, our previous behavioral data support the role of TRPA1 receptors in cold detection, as intraplantar injection of CA in rats resulted in enhanced avoidance of a cold surface (temperature preference test) and significantly lowered the withdrawal threshold at 0°C and +5°C (cold plate test), which are phenomena indicative of cold hyperalgesia (Carstens et al., 2010; Tsagareli et al., 2010; Tsagareli, 2011). Here, in the cold plate test, we revealed cold hyperalgesia in the AITC-treated group and did not observe any antinociceptive effects. These results are consistent with the possibility that TRPA1 agonists can enhance cold-evoked gating of TRPA1 channels to increase their cold sensitivity (Story et al., 2003; Karashima et al., 2009).

In conclusions, TRPA1, being a warm sensor, when it is stimulated by various agonists (e.g., AITC, CA, etc.), the resultant sensation is burning pain rather than cold. However, the role of TRPA1 in cold reception and cold pain sensitivity remains controversial. Our recent data support the role of TRPA1 in cold detection, as the TRPA1 agonist CA enhanced cold sensitivity in two behavioral assays. TRPA1 is undoubtedly involved in pain, and TRPA1 agonists enhance sensitivity to heat pain, possibly by indirectly modulating TRPV1 co-expressed with TRPA1 in nociceptors (Nozadze et al., 2016a, b; Nozadze et al., 2019).

6.2. TRPV1 Channel

TRPV1 is activated by pungent *Capsicum* spices that produce as active components capsaicinoids. Capsaicinoids are the amides of a phenolic amine (vanillamine) with medium-sized, mostly branched, fatty acids. Over 12 major capsaicinoids have been characterized from chili pepper, the most abundant one being capsaicin (Nilius, Appendino, 2011).

Exposure to the skin capsaicin evokes a painful burning sensation through the vanilloid TRPV1 receptor, which is also activated by noxious thermal stimuli above 43°C or by an acidic environment of pH 5.4 (Caterina, 2007; Tominaga, 2009). In addition to its sensitivity to various pain inducing stimuli such as capsaicin, heat, and protons, the TRPV1 ion channel has many features that a receptor related to nociception is supposed to have, such as its preferential distribution in the CNS within small-sized to medium-sized spinal dorsal root and trigeminal ganglia neurons, which are believed to serve as nociceptive nerve cells (Carstens et al., 2010). Although TRPV1 was shown to be activated by noxious heat, studies in TRPV1-knockout mice have revealed intact or partly reduced heat sensitivity (Caterina et al., 2000; Davis et al., 2000; Zimmermann et al., 2005). This suggests that TRPV1 cannot alone be responsible for heat nociception in the 42 – 52°C temperature range, and this also applies to all other heat-activated ion channels as far as knockout mice phenotypes are concerned (Hoffmann et al., 2013).

In this study, we examined by behavioral tests whether capsaicin affect sensitivity to thermal and mechanical stimuli in male rats. We further hypothesized that intraplantar injection of various doses of capsaicin would induce hyper- or hypo-sensitivities to 30 and 15°C temperatures (thermal preference test) and to +5, 0, and −5°C temperatures (cold plate test).

The hindpaw-injected capsaicin yielded a decrease in the withdrawal latency (Figure 21A). The 0.1, 0.2, 0.3, and 0.4% capsaicin-treated groups were all significantly different from the vehicle-treated group ($P < 0.001$), but not from each other. There were some mirror-image effects on the contralateral hindpaw, which were not significantly different among all concentrations, but were significantly different compared with the vehicle group ($P < 0.01$) (Figure 21B).

The von Frey mechanical pressure test showed that for the ipsilateral (treated) hindpaw, the 0.1 – 0.4% capsaicin groups were significantly different from the vehicle group (indicating allodynia, Figure 21C), but not from each other. For the contralateral hindpaw, there were some mirror-image effects, especially for the 0.4% capsaicin concentration ($P < 0.01$) (Figure 21D).

The two-temperature preference test revealed that on 30 vs. 15°C plates, rats treated with a relatively high (0.3%) concentration exhibited significant preference for the colder (15°C) plate compared with vehicle-treated rats ($P < 0.001$) – that is, the animals significantly avoided the warmer (30°C) plate (Figure 22). Another group of rats treated with the highest (0.4%) concentration demonstrated an opposite effect, similar to the effect on the control group – that is, a significant preference for the warmer (30°C) plate (P < 0.001). Two groups of rats treated with lower (0.1 and 0.2%) concentrations of capsaicin exhibited no significant preference for the warmer or colder plate, although the lowest (0.1%) concentration produced a strong preference for the colder (15°C) plate, compared with vehicle treated rats, and this difference across treatments (vehicle *vs.* 0.1% group to 15°C plate) was significant (P < 0.05) (Figure 22).

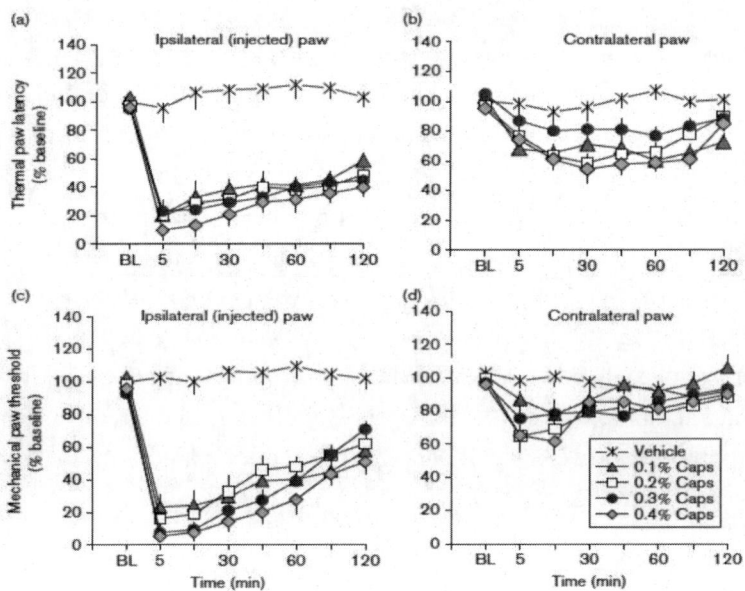

Figure 21. (A) Thermal paw withdrawal latency *vs.* time after ipsilateral intraplantar injection of vehicle (control) and capsaicin at each concentration indicated. There was a significant effect of all concentrations of capsaicin relative to vehicle (P < 0.001). (B) as in (A) for the paw contralateral to AITC injection. (C) as in (A) for Von Frey mechanically evoked withdrawal of the injected paw. There was a significant effect of all capsaicin concentrations relative to vehicle (P < 0.001). (D) as in (C) for the contralateral paw. Note there are some mirror effects for paw withdrawal latency (B) and for mechanical threshold latency (D) (P < 0.01). N = 6/group. BL, pre-injection baseline (adapted from Nozadze et al., 2016b).

We found almost similar results for capsaicin injections (Figure 23) and AITC injections (Figure 20) with the cold plate test, except at −5°C, at which the difference between the vehicle-treated (tween + saline) and capsaicin-treated groups was less significant (Figure 23C) than with AITC injections (Figure 20C).

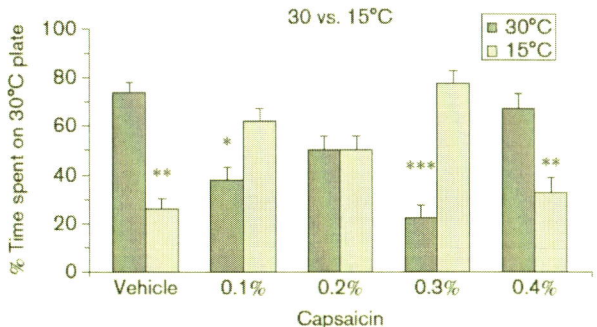

Figure 22. Biphasic effects of capsaicin on thermal preference. The figure shows the % of time for which rats stood on the 30°C vs. 15°C plate; n =12/group. At a higher (0.3%) capsaicin concentration, rats spent significantly more time on the 15°C plate (light gray bars) compared with the 30°C plate (dark gray bars; $P < 0.001$), indicating cold hypo-sensitivity. However, at the highest (0.4%) capsaicin concentration, rats spent significantly more time on the 30°C plate (dark gray bars) compared with the 15°C plate (light gray bars; $P < 0.01$), indicating cold hypersensitivity. At the lower (0.1%, and 0.2%) concentrations, there were no significant differences between the warm (30°C, dark gray bars) and cold (15°C, light gray bars) plates. $^*P < 0.05$; $^{**}P < 0.01$; $^{***}P < 0.001$ (adapted from Nozadze et al., 2016b).

The obtained data showed that intraplantar capsaicin dose-depending increased the magnitude and duration of heat hyperalgesia. Capsaicin at higher concentrations may also trigger central sensitization, leading to the observed reduction in the withdrawal latency for the contralateral paw (Figure 21B and D). With regard to capsaicin-induced effects in human experiments, application of capsaicin on the tongue significantly enhanced heat pain but not cold pain (Albin et al., 2008). This finding is consistent with prior psychophysical studies showing that intradermal capsaicin enhanced the heat pain intensity within a small region around the injection site for up to 2 h (LaMotte et al., 1991, 1992; Torebjork et al., 1992). The

TRPV1 channels that are sensitive to capsaicin respond to temperatures above the pain threshold (Caterina, 2007). The results presented might thus be explained by a capsaicin-induced enhancement of thermal gating of TRPV1 expressed in polymodal nociceptors mediating thermal pain sensation (Carstens et al., 2007; Albin et al., 2008). However, we do not exclude the possibility that, in the two-temperature preference tests, the rats prefer the cool side because they are seeking alleviation of the 'burning' pain (De Felice et al., 2013; Navratilova et al., 2012, 2013) that can be specified on the modulation of TRPV1 channel sensitivity.

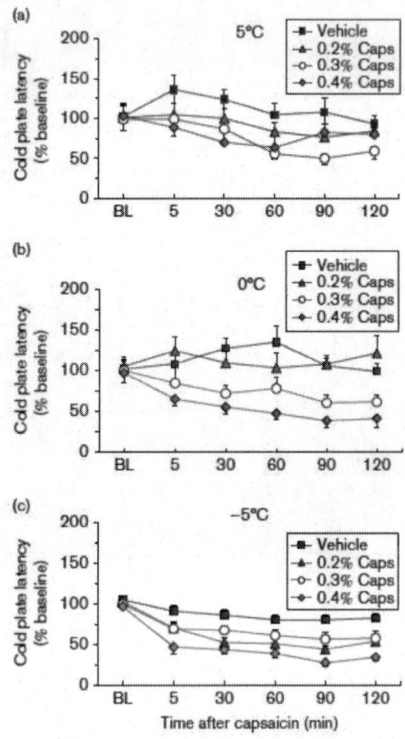

Figure 23. Intraplantar capsaicin injections produce dose-dependent cold hyperalgesia at temperatures of (A) +5°C, (B) 0°C, and (C) −5°C, respectively. The figure shows the change in cold plate latency (% of pre-capsaicin baseline) *vs.* time after topical capsaicin application at the concentrations indicated; n = 12/group. Note the capsaicin effect appears to grow over time, consistent with it diffusing more slowly through the skin (adapted from Nozadze et al., 2016b).

Hoffmann et al. (2013) have recently shown that TRPA1 and TRPV1 channels contribute to thermal nociception, and that both TRPA1 and TRPV1 null mice presented behavioral deficits in heat sensitivity. Furthermore, TRPV1-knockout mice showed both reduced behavioral heat sensitivity and heat-induced CGRP release. They confirmed that TRPV1 is not the sole noxious heat sensor and that other contributors must be expressed in the TRPV1 lineage (Mishra et al., 2011; Hoffmann et al., 2013). These authors hypothesized a possible synergistic or conditional relationship between TRPA1 and TRPV1 receptors involved in the response of cutaneous nociceptors to noxious heat, and direct as well as indirect interactions between the two receptor channels can be considered (Hoffmann et al., 2013).

Recently, the molecular mechanisms underlying TRPV1 and TRPA1 interactions in nociceptive neurons have been elucidated. Particularly, Spahn et al. (2014) found a sensitization of TRPV1 after TRPA1 stimulation with mustard oil in a calcium and cAMP/PKA-dependent manner. TRPA1 stimulation enhanced TRPV1 phosphorylation through the putative PKA phosphorylation site, serine 116. They also detected a calcium-sensitive increase in TRPV1 activity after TRPA1 activation in the DRG neurons. Overall, this study showed sensitization of TRPV1 through activation of TRPA1, which involves adenylyl cyclase, increased cAMP, subsequent translocation and activation of protein kinase A (PKA), and phosphorylation of TRPV1 at PKA phosphorylation residues (Spahn et al., 2014). In other experiments, Fischer et al. (2014) found evidence of a TRPV1–TRPA1 interaction that is predominantly calcium dependent, and it has been suggested that the two proteins might form a heteromeric channel. Thus, these data provide some molecular evidence that supports our behavioral results.

In summary, TRPV1, being a heat sensor, responds to its agonist capsaicin, which elicits corresponding heating sensations and hence lowers the threshold and enhances heat-evoked gating of TRPV1 (Nozadze et al., 2016a, b; Tsagareli, 2013, 2015).

6.3. TRPM8 Channel

The cold- and menthol-sensitive TRPM8 receptor is a non-selective cation channel, with a certain preference for Ca^{2+} permeation. It shows multimodal gating which is activated by cold (< 28°C), membrane depolarization, different cooling compounds such as menthol and icilin, among others, and changes in extracellular osmolality. While phosphoinositides PIP2 is a key regulator of channel gating, testosterone, artemin, and Pirt (phosphoinositides interacting regulator of TRP) protein have been proposed as endogenous ligands of TRPM8. These channels are highly expressed in peripheral sensory neurons (Aδ and C fiber afferents), and also on deep visceral afferents in prostate, broncho-pulmonary tissue, bladder, and the urogenital tract (González-Muñiz et al., 2019).

Menthol, a natural product of the peppermint plant *Mentha x piperita* (Lamiaceae), is a monoterpene which is widely used as a natural product in cosmetics, a flavoring agent, and as an intermediate in the production of other compounds. Various extracts from peppermint contain menthol as a major active constituent and have been used for centuries as traditional medicines for a number of ailments including infections, insomnia, and irritable bowel syndrome as well as an insect repellent (Farco, Grundmann, 2013; Journigan, Zaveri, 2013). Moreover, clinical studies have reported that topical menthol appears safe and effective in treating a variety of painful conditions, including musculoskeletal pain, sports injuries, neuropathic pain and migraine (Pergolizzi et al., 2018). In addition, menthol is also commonly used in food additives and has broad industrial use in oral hygiene and other applications.

Menthol applied to the skin elicits cooling and tingling sensations, and has some anesthetic and *k*-opioid-mediated antinociceptive properties in the mouse hot-plate test (Galeoti et al., 2001, 2002). Menthol has long been known to induce or enhance cooling *via* an interaction with peripheral cold receptors (Hensel, Zotterman, 1951) that express the TRPM8 channel (McKemy et al., 2002; Nealen et al., 2003; Peier et al., 2002). TRPM8 is activated by temperatures below 28°C as well as by menthol and other cooling agents, and knockout mice lacking TRPM8 exhibit decreased sensitivity to cold surfaces that are normally avoided (Bautista et al., 2007; Colburn et al., 2007; Dhaka et al., 2007). Menthol enhances cooling-evoked gating of TRPM8 transfected in cell lines (Malkia et al., 2007; Rohacs et al., 2005; Voets et al., 2004) and naturally expressed in the DRG and TG cells (Madrid et al., 2006; Reid et al., 2002). Higher concentrations of menthol enhance cold pain in human skin (Hatem et al., 2006; Wasner et al., 2008) and oral mucosa, and enhance cold avoidance in rats assessed using an operant facial thermal stimulation paradigm (Rossi et al., 2006). Sensory neurons expressing TRPM8 project to superficial laminae of the spinal cord dorsal horn (Dhaka et al., 2008; Wrigley et al., 2009) which contain cold-sensitive neurons that project in the spino-thalamic tract (Craig, Dostrovsky, 2001). Responses of nociceptive neurons in superficial laminae of trigeminal subnucleus caudalis (Vc) to lingual cooling are enhanced by menthol (Zanotto et al., 2007).

The apparently opposing effects of menthol on the perception of heat and cold pain prompted the present study. We wished to systematically investigate and compare the modulatory effects of topical menthol on thermal (hot and cold) and mechanical pressure sensitivity in rats using an array of behavioral tests (Klein et al., 2010).

6.3.1. Thermal Paw Withdrawal (Hargreaves) Test

The hindpaw receiving topical menthol exhibited a concentration-dependent increase in withdrawal latency (Figure 24A). The 40%, 10% and 1% menthol treatment groups were significantly different from vehicles, and the 40% group was significantly different from all other concentrations (Figure 24A, *$p < 0.01$, repeated measures ANOVA). The 10% menthol group was not significantly different from 1% menthol ($p = 0.07$), and the 0.1% and 0.01% groups did not differ significantly from vehicle. There was an apparent mirror-image effect, in that for the contralateral hindpaw, where the 40% menthol group was significantly different from all other concentrations (*$p < 0.01$) which were not significantly different from vehicle (Figure 24B). Figure 24C and D show data for the vehicle controls.

For the treated hindpaw, there was a significant difference between vehicle groups (*p < 0.01), with 50% ethanol treatment resulting in a significant reduction in thresholds (Figure 24C; *p < 0.01). For the contralateral hindpaw, there was no significant difference between ethanol concentrations, both of which were ineffective (Figure 24D).

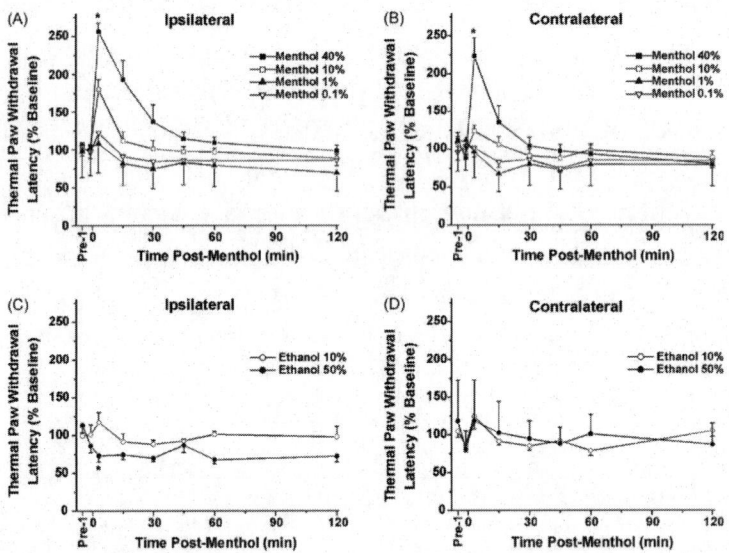

Figure 24. Concentration-dependent antinociceptive effects of menthol on thermal hindpaw withdrawal latency. (A) Thermal paw withdrawal latency (Hargreaves) test: ipsilateral hindpaw. The hindpaw receiving topical menthol exhibited a concentration dependent increase in withdrawal latency (analgesia). Groups of animals tested at concentrations of 40%, 10% and 1% menthol were significantly different from vehicles. 40% menthol was different from all other concentrations (*p < 0.01, repeated measures ANOVA), while 10% menthol was not different from 1% menthol (p = 0.07). Data for 0.01% menthol are similar to 0.1% menthol group and omitted for clarity. Error bars: SEM; n = 8/group. The stimulus was cutoff at 20 s if there was no withdrawal. (B) Contralateral hindpaw. There was a weak mirror-image effect. The 40% menthol group was significantly different from all other concentrations (*p < 0.01), which were not different from vehicle (0.01% menthol omitted). (C) Vehicle controls: ipsilateral hindpaw. There was a significant difference between groups (*p < 0.01). (D) Vehicle controls: contralateral hindpaw. There was no significant difference between ethanol concentrations, both of which were ineffective (reproduced from Klein et al., 2010).

6.3.2. Mechanical Paw Withdrawal (von Frey) Test

For the ipsilateral (treated) hindpaw, the 0.1 – 10% menthol groups were not significantly different from vehicle (Figure 25A). Only the 40% menthol group was significantly different from all other groups (*p < 0.05, repeated-measures ANOVA) indicating allodynia. For the contralateral hindpaw, none of the menthol concentration groups were significantly different from the vehicles (Figure 26B). Figure 25C and D shows data with vehicle controls. There was no significant difference between vehicle groups for the ipsilateral or contralateral hindpaws.

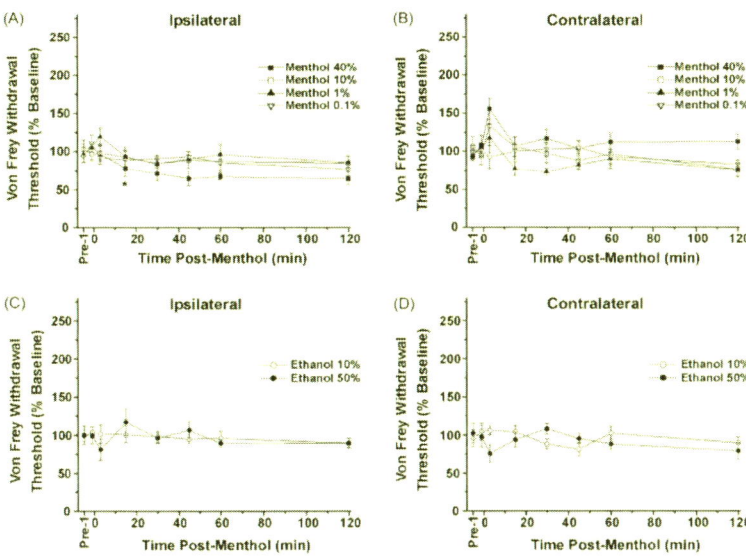

Figure 25. Mechanical paw withdrawal latencies and the lack of concentration-dependent effect of menthol. (A) Von Frey paw withdrawal threshold: ipsilateral hindpaw. The 0.1 – 10% menthol groups were not significantly different from vehicle (10% ethanol). Only the 40% menthol group was significantly different from all other groups (*p < 0.05, repeated-measures ANOVA) indicating allodynia. Data for 0.01% menthol are similar to 0.1% menthol treated groups and omitted for clarity, n = 8/ group. (B) Contralateral hindpaw. None of the menthol concentration groups were significantly different from the vehicles (0.01% menthol omitted). (C) Vehicle controls: ipsilateral hindpaw. There was no significant difference between 10% + 1% Tween-80 and 50% ethanol + 5% Tween-80 vehicle groups. (D) Vehicle controls: contralateral hindpaw. No significant difference between ethanol concentrations (reproduced from Klein et al., 2010).

6.3.3. Two-Temperature Preference Test

Naïve animals significantly avoided temperatures below 30°C and above 35°C (Figure 26A). Untreated (naïve) rats exhibited no preference for either surface when they were both set at 30°C, indicating an absence of positional preference (Figure 26A). Naïve rats did not show a preference for 35°C *vs.* 30°C, indicating that their preferred temperature lies within this range or possibly 1 – 2°C higher or lower since we only tested temperature differentials of 5°C. When one of the plates was set to a temperature of 25°C and lower, or 40°C and higher, there was a temperature-dependent decrease in the percent time spent on the colder or hotter plate which was significantly different compared to time spent on the 30°C plate (Figure 26A). These temperature preference results are similar to previous data showing that mice on a thermal gradient spent the majority of time in the 30 – 36°C range (Moqrich et al., 2005). Figure 26B plots the mean number of times rats crossed between the two plates. The maximum number of crossings was observed when both plates were set at 30°C and decreased when the non-neutral plate was set at higher or lower temperatures (Figure 26B). The greatest decline in number of crossings, as well as time spent on the non-neutral plate, was observed for the largest temperature differences, i.e., 0°C and 50°C *vs.* 30°C (Figure 26A and B).

Menthol had a biphasic effect on temperature preference. For formal testing, one plate was set at 30°C and the other at 15°C or 20°C in a counterbalanced design. These temperature differentials were chosen because the degree of avoidance of both temperatures (20–30%; see Figure 26A) was intermediate compared to warmer or colder temperatures, thus allowing for menthol induced shifts in either direction (i.e., avoiding floor or ceiling effects). In the 15°C *vs.* 30°C preference test (Figure 27A), treatment with high menthol concentrations (10% and 40%; vertically-striped and open bars) resulted in rats spending a significantly ($p < 0.05$) lower proportion of time on the 30°C plate compared to vehicle controls (diagonally-hatched and dark gray bars), indicating cold hyposensitivity.

At lower concentrations (0.01–1%; horizontally-striped, light gray and stippled bars in Figure 27A), rats spent significantly more time on the 30°C plate (p < 0.05) indicating cold hypersensitivity. At the lowest menthol concentration (0.01%) there was a significant decline in the number of plate crossings (9.7 ± 1.3) when compared to naïve or vehicle treated animals (16.7 ± 2.3 and 15.4 ± 1.8, respectively, p < 0.05 for both). There were no significant differences among control groups, with the 10% ethanol, 50% ethanol, and untreated naïve groups exhibiting preferences for the 30°C plate of 79.9 ± 6.6%, 77 ± 7.0%, and 80.3 ± 6.0%, respectively (Figure 27A).

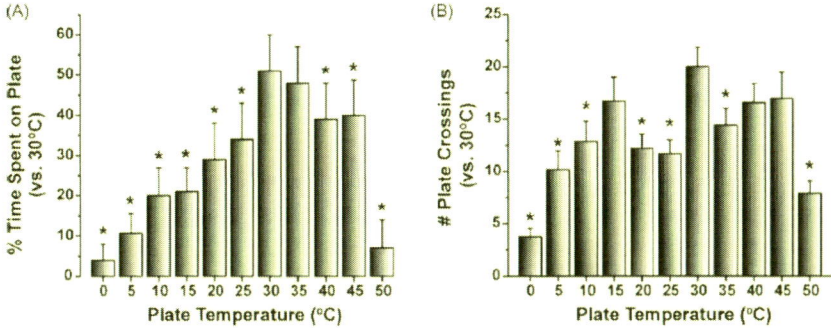

Figure 26. Two-temperature preference test. Rats were placed on one of two adjacent thermoelectric plates whose temperatures could be set independently (range − 5°C to > 50°C). One plate was set at 30°C and the other at a warmer or colder temperature in 5°C increments in a counterbalanced design. The rat was free to move from one surface to the other. (A) The graph plots the mean percentage of time naive rats spent on the warmer or colder plate relative to the thermo-neutral (30°C) plate over a 20 min period. Rats significantly avoided temperatures < 30°C and > 35°C. *p < 0.05, paired t-test. (B) as in (A) for mean number of crossings between the thermo-neutral (30°C) and warmer/ cooler plate over a 20 min period. Error bars: SEM. Naive animals, *p < 0.05, unpaired t-test; n = 16/group for 5 − 45°C; n = 8/group for 0°C and 50°C) (reproduced from Klein et al., 2010).

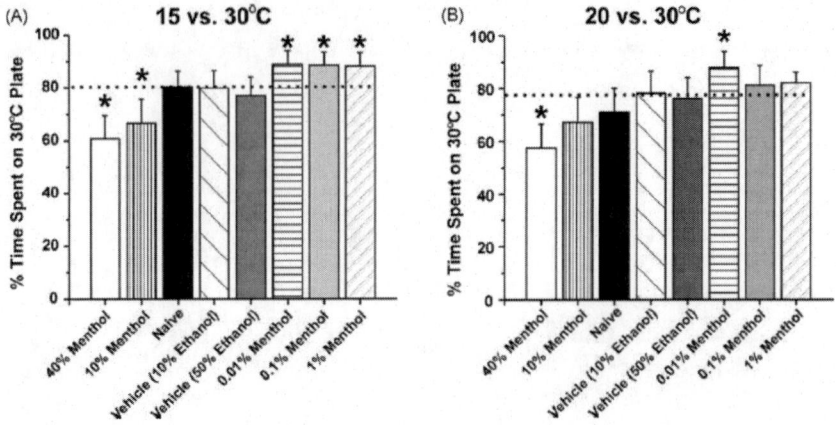

Figure 27. Biphasic effects of menthol on thermal preference. (A) Graph plots % time rat stood on 30°C vs. 15°C plate. Horizontal dashed line indicates that naive and vehicle treated rats avoided the colder plate ~ 80% of the time as a reference. At high (40%, 10%; open and vertically-striped bars) menthol concentrations rats spent significantly less time on the 30°C plate compared to vehicle controls (diagonally-hatched and dark gray bars) ($p < 0.05$), indicating cold hyposensitivity. At lower concentrations (0.01 – 1%; horizontally-striped, light gray and stippled bars), rats spent significantly more time on the 30°C plate ($p < 0.05$) indicating cold hypersensitivity. * Significantly different vs. vehicle ($p < 0.05$, n = 16/group). (B) as in (A) for 20°C vs. 30°C preference test. Note significant cold hyposensitivity with 40% menthol and significant hypersensitivity at the lowest menthol concentration of 0.01%. * Significantly different vs. vehicle ($p < 0.05$, n = 16/group) (reproduced from Klein et al., 2010).

In the 20°C vs. 30°C preference test (Figure 27B), a similar biphasic effect was noted, with the highest menthol concentration (40%) resulting in significantly less time, and the lowest concentration (0.01%) resulting in significantly more time, spent on the 30°C plate. At the lowest concentration (0.01%) of menthol, there was also a significant decline in the number of plate crossings (8.25 ± 0.8) when compared to naïve or vehicle treated animals (16.7 ± 2.3 and 15.4 ± 1.8, respectively, $p < 0.05$). At the highest menthol concentration tested (40%) there was a significant increase in the number of plate crossings compared to naïve or vehicle treated animals (18.4 ± 1.8 vs. 12.2 ± 1.3 and 11.6 ± 1.5, respectively, $p < 0.05$). There were no significant differences among control groups, with the 10% ethanol, 50% ethanol, and untreated naïve groups exhibiting preferences for the 30°C plate of 78.1 ± 8.4%, 76.1 ± 7.9%, and 71.0 ± 9.0%, respectively (Figure 27B).

6.3.4. Cold Plate Test

The mean baseline latencies on the −5.0°C and 0°C cold plates were 26.4 ± 0.4 s and 105.9 ± 2.3 s, respectively. All animals displayed a nocifensive response within the 150 s cutoff time on the −5°C cold plate and 75% did on the 0°C cold plate. In the −5°C cold plate test (Figure 28A), the 40% menthol group was significantly different from all other concentrations (*$p < 0.001$, repeated-measures ANOVA) which did not differ from each other or vehicle. This suggests that the high menthol concentration induced cold hypo-algesia. Figure 28B shows the vehicle groups, with a significant difference between 10% and 50% ethanol treatments (*$p < 0.001$, ANOVA). The ~20% reduction in cold plate latency seen at 3 min post-10% ethanol was comparable to the latency reductions in the 0.1–10% menthol groups at the same 3-min time point (Figure 28A). In the 0°C cold plate test, there was no significant difference among menthol concentration groups (Figure 28C) and no significant difference between vehicle (ethanol) concentrations (Figure 28D).

The main obtained findings may be summarized as follows. Menthol increased paw withdrawal latencies to noxious heat in a concentration-dependent manner, indicating an antinociceptive effect. The highest menthol concentration also significantly increased cold plate latencies, consistent with antinociception. However, lower menthol concentrations did not significantly affect nocifensive latencies in the cold plate test, indicating that menthol more effectively suppresses heat compared to cold pain. These effects are unlikely to be explained by a local anesthetic effect, since the highest menthol concentration increased mechano-sensitivity (allodynia). Menthol had a biphasic effect on innocuous cold sensitivity, with high menthol concentrations reducing and low menthol concentrations enhancing avoidance of cooler surfaces. These findings are discussed in relation to possible underlying neural mechanisms.

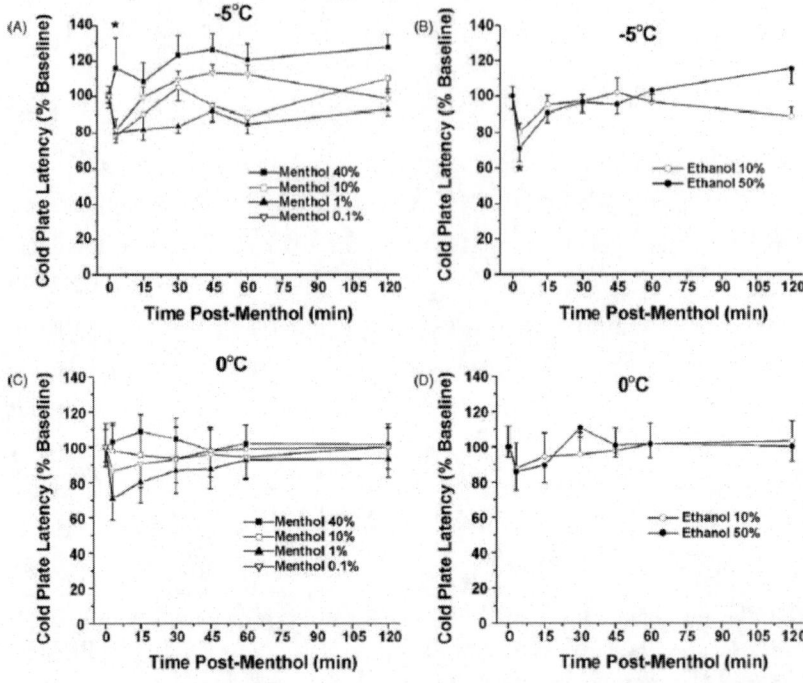

Figure 28. Antinociceptive effects of 40% menthol in −5°C cold plate test. (A) Cold plate −5°C: menthol. Graph plots change in cold plate latency (% of pre-menthol baseline) *vs.* time after topical menthol application at indicated concentrations. Mean baseline latency was 26.4 s; animals were removed at 150 s (cutoff) if they did not respond. Rats were tested 3 min after menthol or vehicle, and again at 15-min intervals out to 60 min and once more at 120 min. The 40% menthol group was significantly different from all other concentrations (* $p < 0.001$, repeated-measures ANOVA) which did not differ from each other or vehicle. This indicates a cold antinociceptive effect of 40% menthol. Error bars: SEM, n = 8/group. (B) Cold plate − 5°C: vehicle. There was a significant difference between 10% and 50% ethanol (* $p < 0.001$, ANOVA). The ~ 20% reduction in cold plate latency seen at 3 min post-10% ethanol was comparable to the latency reductions in the 0.1 − 10% menthol groups at the same 3-min time point (A). (C) Cold plate 0°C: menthol. Mean baseline latency was 105.9 s. There was no significant difference among menthol concentration groups. (D) Cold plate 0°C: vehicle. There was no significant difference between ethanol concentrations (reproduced from Klein et al., 2010).

6.3.5. Antinociception

Topical balms and other over-the-counter products for pain relief often contain menthol concentrations of 5–16% or even higher. The present data indicate that topical application of menthol in this concentration range is antinociceptive for heat pain, and also for cold pain at the highest menthol concentration. These findings are consistent with previous studies showing menthol suppression of heat pain (Albin et al., 2008; Green, 1992, 2005) and capsaicin irritancy (Green, McAuliffe, 2000). The mechanism of menthol's antinociceptive effect is not certain although many menthol-sensitive DRG (Reid et al., 2002) and Vc neurons (Zanotto et al., 2007) respond to capsaicin and other noxious stimuli and innocuous cooling can elicit nociceptive sensations (Green, Pope, 2003). Topical paw application of menthol has also been reported to reverse behavioral reflex sensitization to noxious heat and mechanical stimulation in the rat chronic constriction injury model of neuropathic pain (Proudfoot et al., 2006). These data suggest that menthol-sensitive primary afferent fibers can inhibit nociceptive pathways. A peripheral mechanism could involve menthol inhibition of nociceptors, possibly by blocking TRPA1 expressed in nociceptive nerve endings (Karashima et al., 2007; Macpherson et al., 2006). Another mechanism is menthol activation of cold receptors that centrally inhibit spinal nociceptive neurons. A third possibility is that menthol engages supra-segmental or supra-spinal circuits to result in descending inhibition of spinal nociceptive neurons.

Topical application of menthol elicits oral irritation (Cliff, Green, 1994, 1996; Dessirier et al., 2001) and cutaneous cold pain (Green, Shoen, 2007; Wasner et al., 2004) and directly excites many cold-sensitive nociceptive Vc neurons (Zanotto et al., 2007, 2008). Unilateral menthol induced a weaker mirror-image antinociceptive effect in the contralateral hindpaw (Figure 24B), suggesting the involvement of hetero-segmentally-organized antinociceptive circuits that exerted a depressant effect on nociceptive neurons bilaterally. This would be akin to counter-irritation, which was demonstrated in a human psychophysical study in which chemically-evoked irritation on one arm was suppressed by a stronger irritant stimulus delivered to the opposite side (Green, 1991).

There was a small but significant reduction in paw withdrawal latency following the 50% ethanol vehicle (Figure 24C) that might be attributed to ethanol sensitization of TRPV1 expressed in nociceptors responsive to noxious heating (Vetter et al., 2008). This effect might have slightly reduced the analgesic effect of high menthol concentrations (10% and 40%) that were dissolved in 50% ethanol.

6.3.6. Cold Sensitivity

The present data revealed a biphasic effect of menthol on innocuous cold sensitivity, with high menthol concentrations reducing and low menthol concentrations enhancing the avoidance of cold temperatures. We tested preference for a thermo-neutral (30°C) surface *vs.* 15°C and 20°C surfaces, since the latter cold temperatures are avoided about 70–80% of the time by naïve rats (Figure 26A). Rats receiving high (10% and 40%) menthol concentrations avoided the 15°C and 20°C surfaces to a significantly lesser degree, indicating indifference to the colder surface that might reflect cold hypo-algesia. They also exhibited a high number of plate crossings, comparable to naïve animals tested with both plates set at 30°C. We reasoned that when animals initially stood on the colder plate and perceived it to be aversive they tended to subsequently avoid it thus resulting in fewer plate crossings. However, while there was a systematic decline in time spent on surfaces having progressively colder or hotter temperatures (Figure 26A), the relationship of plate crossings to temperature difference was more variable (Figure 26B) suggesting that plate crossings are a less sensitive measure of cold or heat aversion. The indifference to cold temperatures might be attributed to a peripheral desensitization of TRPM8 and/or TRPA1 by high menthol concentrations (Abe et al., 2006) or a central inhibitory effect, as described above.

In contrast, low concentrations of menthol (0.01–1%) significantly increased avoidance of the 15°C and 20°C surfaces and reduced the number of plate crossings (Fugure 27A, B). These results might reflect cold allodynia, or they may indicate an increase in sensitivity to innocuous cold that is aversive to the animal but not actually painful. It is interesting that the decrease in cold sensitivity observed in TRPM8 knockout mice appears to disappear below 10°C (Bautista et al., 2007), a temperature that is often reported to be painful. This would be consistent with a role for TRPM8 in innocuous cold sensation but not pain. In any event, the increase in cold or cold pain sensitivity may involve menthol enhancement of thermal gating of TRPM8 expressed in cold receptors and/or nociceptors.

In the cold plate test, the effect of low menthol concentrations on nocifensive latency did not differ from vehicle (Figure 28A, C), which we interpret as an absence of cold hyperalgesia. Previous human studies have generally reported cold hyperalgesia and/or a decrease in cold pain threshold following topical menthol application (Green, 1992; Hatem et al., 2006; Namer et al., 2005; Wasner et al., 2004). However, thresholds for cold pain are more variable compared to heat pain (Neddermeyer et al., 2008), so that overall subjective ratings of cold pain might include a component of innocuous cold that is also enhanced by menthol.

The same high concentration of menthol (40%) that induced cold hyperalgesia in human studies reduced cold sensitivity in the rat thermal preference and −5°C cold plate tests (Figure 28A, C). The observed cold hyposensitivity is noteworthy given that vehicle (50% ethanol) had the opposite effect of reducing nocifensive response latency initially (Figure 28B). The latter effect might be explained by an initial and brief evaporative cooling effect of ethanol that summed with physical skin cooling during initial hindpaw contact with the cold plate.

The opposing effects of 40% menthol in human (cold hyperalgesia) *vs.* rat (cold hyposensitivity) might be explained by allometric scaling and/or differences in dermal diffusion. We presently applied menthol to both ventral hindpaws, which constitutes a substantially larger percentage of the overall body surface area of a rat, as compared to the restricted region of volar forearm skin treated with menthol in the human studies (Hatem et al., 2006; Namer et al., 2005; Wasner et al., 2004). Moreover, dermal absorption of chemicals, which is a function of the total area of application and concentration (Magnusson et al., 2001) is greater in the rat *vs.* human (Walters, Roberts, 1993). For example, permeation of L-menthol through the stratum corneum was four times higher in the hairless rat compared to human skin (Sugibayashi et al., 1995). We speculate that the relative area of chemical stimulation and species differences in skin permeation explain why several-fold lower concentrations of menthol induced cold hypersensitivity in rats compared to the human studies.

6.3.7. Mechanosensitivity

Menthol had very little effect on mechanical paw withdrawal thresholds, consistent with a general lack of effect of menthol on mechanosensitivity in human skin (Hatem et al., 2006; Namer et al., 2005). This differs from previous reports of a significant, dose-dependent allodynia following intraplantar injection of the TRPV1 agonist capsaicin (Gilchrist et al., 1996) or the TRPA1 agonist cinnamaldehyde (Tsagareli et al., 2010) in rats. We presently observed a mild allodynia the highest menthol concentration tested, which mitigates against a local anesthetic action of menthol.

6.3.8. Thermo-TRPs and TRPM8 in Cold Transduction

The mechanism of cold transduction has been elusive, and despite the discovery of thermo-TRPs it remains a complex issue (Jordt et al., 2003; Reid, 2005; Reid, Flonta, 2001). The behavioral effects of topical menthol application seen in this study are consistent with effects on TRPM8. An important role for TRPM8 becomes apparent when this channel is missing. TRPM8 null mice exhibit a deficit in cold avoidance and lower incidence of cold-sensitive afferent fibers (Bautista et al., 2007; Colburn et al., 2007; Dhaka et al., 2007). However, the authors cannot rule out the possibility that menthol may interact with other channels expressed in sensory neu+rons (Galeotti et al., 2002; Macpherson et al., 2006; Munns et al., 2007). Recent observations suggest that cold sensation likely involves multiple channels, including potassium channels (Kang et al., 2005; Noel et al., 2009; Reid, Flonta, 2001), in transducing and modulating the transmission of temperature information (Belmonte et al., 2009; Viana et al., 2002).

Overall, presented results are generally consistent with neurophysiological and human psycho-physiological data and support TRPM8 as a potential peripheral target of pain modulation (Carstens et al., 2010; Klein et al., 2010; Tsagareli, 2013, 2015).

Chapter 7

TRP CHANNELS AS THERAPEUTIC AND ANALGESIC TARGETS

Progress in cell biology, genetics, molecular, and systems pharmacology is the driving force behind a current paradigm shift in drug research. This paradigm shift shapes new avenues for advanced treatments that are commonly referred to as 'systems therapeutics'. The latter differ in many ways from current drugs because they target biological networks rather than single transduction pathways, and affect disease processes rather than basic physiological mechanisms (Danhof et al., 2018).

Failures in the current paradigm for drug development have resulted in soaring research and development costs and reduced numbers of new drug approvals. Over 90% of new drug programs fail, the majority terminated at the level of Phase 2/3 clinical trials, largely because of efficacy failures or unexplained toxicity (Marshall et al., 2018).

Nowadays, current pain therapeutics mainly includes non-steroidal anti-inflammatory drugs (NSAIDs) and opioids, but they exhibit limitations in efficacy, unwanted side effects and the problem of drug abuse. To overcome these issues, the discovery of different molecular players within pain pathways could lead to new opportunities for therapeutic intervention. Among other strategies, peptides could be powerful pharmaceutical agents for effective opioid-free medications for pain treatment. One of reliable approaches is to use representatives of non-opioid analgesic peptides acting directly or indirectly on different ion channels and receptors distributed in nociceptive pathways. They include peptides targeting Ca^{2+}, Na^+ and K^+ voltage-gated ion channels, the neuronal nicotinic receptors (nAChR), TRP channels, and different non-opioid GPCRs, like the calcitonin gen-related peptide (CGRP), cannabinoids, bradykinin and neurotensin receptors, among others. Peptides engineered from protein-protein interactions among pain-related receptors and regulatory proteins also led to new therapeutic approaches for pain management. Following some successful examples, already in the clinics or under clinical trials, the improved understanding of pain mechanisms, and the advances in peptide permeation and/or delivery, could afford new analgesic peptides in the near future (Perez de Vega et al., 2008).

Data from animal models, human genetics and of some clinical trials will serve as the final arbiters of utility of TRP channel modulators as therapeutics in channel dysfunction (known as TRP channelopathies) (Dietrich, 2019b; Julius, 2013; Nilius, Owsianik G, 2010; Raouf et al., 2010). Current evidence indicates that channelopathies contribute to the development and/or progression of the symptoms of many diseases. Among them are inherited pain syndrome and neuropathic pain, itch, multiple kidney disease, skeletal disorders, overactive bladder, asthma and respiratory system, and of anxiety disorders. Therefore, TRP ion channels could be therapeutic targets that are amenable to blockade by small molecules (Dietrich, 2019a; Moran, 2018; Tsagareli, 2013).

Several hereditary diseases caused by defects in genes encoding TRP channels have been described (see Nilius, Szallasi, 2014; Moran, 2018). Given the pivotal role that TRP channels are thought to play in nociceptive transduction (Julius, 2013; Moran et al., 2011), it is somewhat unexpected that so far, only an obscure human painful condition, Familial Episodic Pain Syndrome, has been linked to a TRP channelopathy, namely, a gain-of-function mutation in TRPA1 channels. That said, multiple polymorphisms in TRP channels have been linked to pain susceptibility (Moran, Szallasi, 2018).

The discovery of novel analgesic drug targets is an active research topic owing to insufficient treatment options for persisting pain. Modulators of thermo-TRP channels, in particular TRPV1, TRPV2, TRPM8 and TRPA1, have reached clinical development. This requires access for TRP channels and the effects of specific modulators in humans. This is currently possible *via* (1) the study of TRP channel function in human-derived cell lines, (2) immune-histochemical visualization of TRP channel expression in human tissues, (3) human experimental pain models employing sensitization by means of topical application of TRP channel activators including capsaicin (TRPV1), menthol (TRPM8), mustard oil and cinnamaldehyde (TRPA1), and (4) the study of phenotypic consequences of human TRP gene variants (Weyer-Menkhoff, Lötsch, 2018).

Despite the striking progress in our understanding of TRP channel functions for therapies (see Moran, 2018), some inherent problems persist. Activation or inhibition of a TRP channel may be beneficial in one organ and, at the same time, may induce unacceptable adverse effects in another. Indeed, the clinical development of first-generation TRPV1 antagonists was halted because they caused hyperthermia and put patients at risk for scalding injuries by elevating the heat pain threshold. Hopefully, drug discovery companies will be able to create a new generation of targeted therapies (Kaneko, Szallasi, 2014).

The roles of TRP channels in mediating pathological pain make them potential targets for analgesics. TRP channels are located in nociceptors where pain is generated, and thus the simplest access for analgesics would involve blocking the channels directly. Two main approaches have been proposed to inhibit TRP channels: blocking the channel activity using antagonists and, paradoxically, stimulating the channels by agonists to desensitize them. Drug development studies have focused on TRPV1, TRPV3, TRPM8, and TRPA1, which have been clearly demonstrated to be involved in pathological pain. Currently, at least seven antagonists of TRPV1, two of TRPA1, and one of TRPV3 are being clinically tested by pharmaceutical industries. However, several TRP channels perform dual functions: they act as sensors/detectors of nociceptive signals under normal conditions, which help prevent tissue damage, and they also act as contributors to inflammatory or neuropathic pain under pathological conditions, and this can pose a risk of noxious perception being blunted when the channels' activity is blocked. In accord, in healthy human volunteers, increase in heat pain threshold was observed after the administration of TRPV1 antagonists, including SB-705498, MK-2295, ABT-102, and AZD1386. Hyperthermia is another adverse effect of TRPV1 antagonists; almost all TRPV1 antagonists undergoing clinical development, such as AMG517 (Amgen), ABT-102 (Abbott), and AZD1386 (AstraZeneca), caused hyperthermia in human volunteers that in certain cases lasted for 1 – 4 days, with core body temperatures rising up to 40°C (Dai, 2016; Kaneko, Szallasi, 2014; Moran, Szallasi, 2018; Moran, 2018).

In contrast to the challenges associated with developing antagonists, in the case of agonists that have been applied to locally desensitize TRP channels and have been used clinically, severe adverse effects have not been reported, although the agonists might produce initial irritation or even degeneration of sensory nerves. This notion of a paradoxical use of agonists might be derived from the experience gained with certain herbal remedies used in traditional medicine. For example, preparations or prescriptions containing agonists of TRP channels, such as menthol (TRPM8), cinnamaldehyde (TRPA1), or shogaol (TRPV1), have long been used topically or orally to relieve neuralgia, arthralgia, menstrual pain, and headache in traditional Chinese medicine. Moreover, capsaicin-containing creams, occlusive patches, and liquid formulations have been developed and used for treating chronic painful conditions such as diabetic neuropathy, postherpetic neuralgia, and other painful disorders (Dai, 2016; Moran, Szallasi, 2017; Moran, 2018).

In summary, TRP channels are involved in diverse physiological processes. The number of TRP channel mutations that lead to disease underscores their importance in human biology. Both agonists and antagonists of TRP channels are currently under development as potential therapeutics. Owing to the relatively low sequence homology among family members and the dramatic differences in three-dimensional structure, truly selective modulators are attainable. Because of these disparate structures and the disparate roles these channels fulfill, each family member will need to be evaluated individually – extrapolation from one channel to another will be difficult (Moran, 2018).

The subgroup of thermo-TRPs, have been well established as key players in pain and analgesia. Results of preclinical research involving several different thermo-TRPs, but with a preference for TRPV1, followed by TRPM8 and TRPA1, are increasingly being translated into the development of novel analgesics. Current clinical trials test thermo-TRP agonists as well as thermo-TRP antagonists that target several different TRPs involved in different pain syndromes including neuropathic pain, rheumatic disease, and painful lesions of the spinal cord, inflammatory bowel diseases, fibromyalgia and diabetic neuropathy. In the translational phase of drug development, the access to TRP channels in humans becomes increasingly important. Reported species differences in TRP channel function emphasize the need to study TRPs in the human setting. The function of human TRP channels can be studied *in vitro* using human cell lines and the transfection of human TRPs or *in vivo* by means of applying TRP-specific activators in human experimental pain models. The expression of human TRP channels is directly accessible *via* mRNA quantification or histochemical visualization. Furthermore, human TRP channels are accessible *via* genetics studies associating functional genotypes with pain-related phenotypes (Weyer-Menkhoff, Lötsch, 2018). The present chapter review concludes that the necessary means for human TRP research have been developed to translate TRP-related analgesic drug investigations into human-centered clinical pharmacological trials.

Thus, there is strong experimental and clinical evidence to substantiate TRP channels as appealing drug targets and a number of molecules targeting TRP channels have already advanced to clinical assays. To overcome these challenges, further investigations must be conducted using newly devised approaches, such as the discovery of second-generation antagonists or selective delivery of drugs through topical application or local injection instead of systemic administration. In principle, analgesics that act at nociceptor-specific targets have the advantage of biological specificity and – for some applications – selective delivery *via* topical application, local injection, or inhalation.

Chapter 8

TRP CHANNELS AND NSAIDS

During inflammation, several TRP channels are directly or indirectly activated by inflammatory signaling molecules and micro-environmental changes including heat, oxidative conditions or low pH. In either case, specific TRP isoforms participate in chains of pro- or anti-inflammatory signaling cascades often including activation of transcription factors, protein kinases and phospholipases, which result in signal integration or amplification. In a few cases, their potentials as therapeutic targets for inflammatory conditions like asthma, cystitis, dermatitis, pruritus among other conditions are investigated pre-clinically or clinically. Significant efforts are still devoted to the understanding of the detailed physiological roles played by TRP channels during inflammation (Benemei et al., 2015; De Logu et al., 2016; Dietrich, 2019b; Parenti et al., 2016; Radresa et al., 2013).

Opiates are fantastically effective pain suppressors, but they are plagued by serious side effects (for instance, tolerance, addiction, constipation, and respiratory depression) owing to actions at receptors outside of the pain pathway. Similarly, NSAIDs are effective at treating inflammatory pain, but their use is limited by gastrointestinal, renal, and cardiovascular risk owing to inhibition of cyclooxygenase enzymes throughout the body (Julius, 2013). The TRP channels discussed here may satisfy these criteria because, although they have been described in tissues and cell types outside of the somatosensory system, their expression (whether measured histologically or functionally) is substantially higher in nociceptors.

In particular, based on clinical and preclinical evidence, TRP channels have emerged as potential drug targets for the treatment of osteoarthritis, rheumatoid arthritis, and gout. Conservative management of these diseases are focused primarily on pain alleviation *via* NSAIDs (Tsagareli, Tsiklauri, 2012), corticosteroid injections, physical therapy, and the use of assistive devices. These interventions typically result in moderate improvements in pain and function over the short term, but none of these treatments is ultimately able to halt disease progression. Additionally, long-term use of NSAIDs is associated with the development of gastrointestinal complications and increased risk of cardiovascular events, which is especially concerning given that patients with osteoarthritis are often older and have multiple comorbidities (Galindo et al., 2018).

Pharmacological findings indicate that TRPA1 contributes to inflammatory and neuropathic pain models, including those evoked by formalin, spinal nerve ligation and chemotherapeutics. Therapeutic effects of NSAIDs are mainly attributed to inhibition of prostanoid synthesis by a non-selective, reversible inhibition of both cyclo-oxygenase 1 (COX1) and COX2 (Benemei et al., 2015; Nassini et al., 2014, 2015). NSAID-mediated inhibition of COX results in inhibition of prostaglandins (PGs) synthesis. PGs in turn cause sensitization and enhance pain signals in spinal neuronal circuits. PGs are bioactive compounds of prostanoids, which include prostacyclin and thromboxane. Prostanoids originate from arachidonic acid, which is released intra-cellularly from plasma membrane phospholipids upon tissue damage and inflammation. Constitutively expressed COX-1 and inducible COX-2 convert arachidonic acid to the precursors PGG_2 and PGH_2, from which all prostanoids are generated by tissue-specific synthases (Zeilhofer, Brune, 2013).

A number of arachidonic acid derivatives, including several electrophilic PGs, have been shown to activate TRPA1 channels and covalent modification is thought to be the main mechanism underlying the direct activation of TRPA1 channels by these arachidonic acid derivatives, because the non-electrophilic precursors of these prostaglandins failed to show the same effect (Hu et al., 2010; Maher et al., 2008; Materazzi et al., 2008; Taylor-Clark et al., 2008). However, it is not known whether the overall structure of prostanoids also contribute to channel regulation. It was recently shown that pyrazolone derivatives, such as antipyrine, dipyrone and propyphenazone, selectively inhibited calcium currents in TRPA1-expressing cells and acute behavior responses in mice evoked by channel agonist, AITC (Nassini et al., 2015). In other experiments hesperidin ("vitamin P") demonstrated antinociceptive activity and synergistic response when combined with ketorolac, possibly by involvement of the TRPV1 receptor ion channel, suggesting their clinical potential in pain therapy (Martínez et al., 2011).

Here we report that some commonly used NSAIDs, such as diclofenac, ketorolac, and lornoxicam (xefocam) probably inactivate or desensitize TRPA1 channel, to enable treatment with CA and AITC, and TRPV1 channel to enable treatment with capsaicin (Nozadze et al., 2018; Tsagareli et al., 2018).

8.1. Analgesic and Antihyperalgesic Effects of NSAIDs Pretreatment

When pretreated with vehicle, intraplantar injection of CA (Figure 29A, C), AITC (Figure 30A, C) or capsaicin (Figure 31A, C) resulted in thermal and mechanical hyperalgesia that persisted beyond 2 h (the dashed lines). These hyperalgesic effects are very similar to those reported previously (Nozadze et al., 2016b; Tsagareli et al., 2010). Pretreatment with diclofenac, ketorolac and xefocam resulted in a strong analgesic effect prior to TRP agonist injections (Figures 29A, C; 30A, C; 31A, C, filled squares). Following the injection of CA, AITC and capsaicin, thermal withdrawal latencies and mechanical thresholds dropped, but there was a pronounced shortening of the duration of thermal and mechanical hyperalgesia (Figures 29A, C; 30A, C; 31A, C). There were significant differences between the experimental compared to control groups in the increase of the thermal paw latency and mechanical withdrawal thresholds for each NSAID: diclofenac, t = 14.165, P < 0.001, ketorolac, t = 22.198, P < 0.001, and xefocam, t = 34.144, P < 0.001, respectively.

Twenty min following intraplantar NSAIDs, microinjection of CA caused a significant decrease of thermal paw withdrawal latencies compared to the pre-injection level at time 0 for the groups of diclofenac, t = 9.229, p < 0.001, ketorolac, t = 10.741, p < 0.001, and xefocam, t = 11.884, p < 0.001, at 5 min of post-CA microinjections, respectively (Figure 29A). Similar results were observed for AITC and capsaicin. Particularly, microinjections of AITC induced a significant decrease of thermal paw withdrawal latencies compared to the pre-injection level at time 0 for the groups of diclofenac, t = 4,952, p < 0.01, ketorolac, t = 13, 853, p < 0.001, and xefocam, t = 4,775, p < 0.01, at 5 min of post-AITC microinjections, respectively (Figure 30A). In the same manner, microinjection of TRPV1 agonist capsaicin resulted in a significant decrease of thermal paw withdrawal latencies compared to the pre-injection level at time 0 for the groups of diclofenac, t = 17,339, p < 0.001, ketorolac, t = 9, 666, p < 0.001, and xefocam, t = 16,053, p < 0.001, at 5 min of post-capsaicin microinjections, respectively (Figure 31A). The reduction in thermal paw withdrawal latency following each TRP agonist injection was approximately equivalent to the reduction observed in the groups receiving saline followed by TRP agonists (dashed lines in Figures 29A, 30A, 31A). After approximately 30–40 min, hyperalgesic effects observed in the groups receiving NSAIDs followed by CA, AITC and capsaicin returned to baseline levels (Figures 29A, 30A, 31A). These findings indicate that pretreatment with NSAIDs did not reduce the magnitude of hyperalgesia, but significantly shortened the time-course of hyperalgesia induced by CA, AITC and capsaicin.

Similar effects were observed for the plantar mechanical pressure test. Intraplantar injection of CA, AITC and capsaicin resulted in a reduction in mechanical paw withdrawal thresholds (Figures 29C, 30C, 31C, dashed lines). Injection of NSAIDs followed by saline resulted in an increase in threshold (Figures 29C, 30C, 31C, filled squares). After pretreatment with each NSAID, injection of CA, AITC and capsaicin still lowered withdrawal thresholds significantly (diclofenac: t = 12.02, p < 0.001; ketorolac: t = 9.749, p < 0.001; and xefocam: t = 11.136, p < 0.001, Figures 29C, 30C, 31C). The reduction in mechanical paw withdrawal threshold following each TRP agonist injection was equivalent between saline and NSAIDs pretreatment groups. However, the duration of the hyperalgesic effect was shorter in the NSAIDs pretreatment groups.

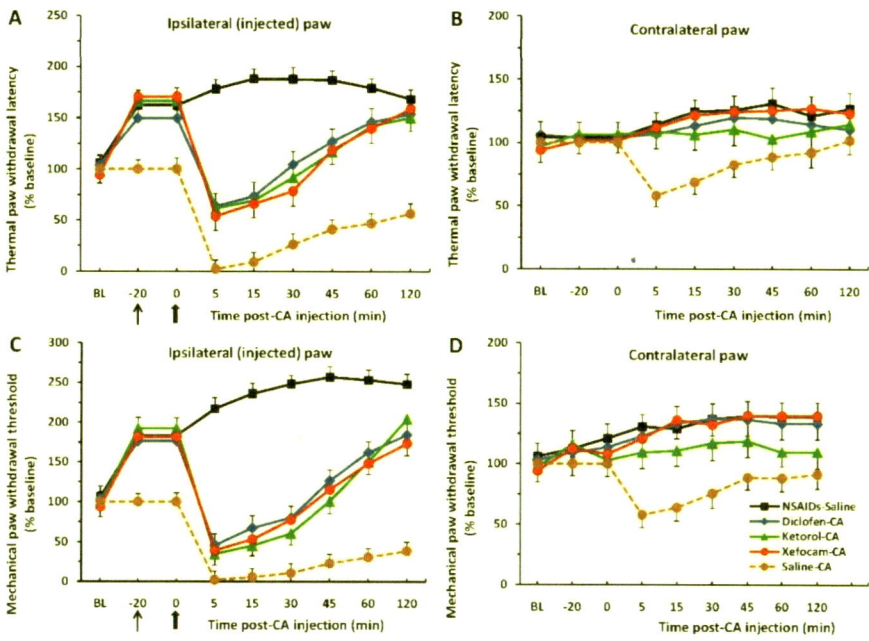

Figure 29. (A) Dynamics of the thermal paw withdrawal latency after NSAIDs pretreatment following ipsilateral intraplantar injection of CA and vehicle. There are significant effects in CA groups vs. vehicle control and contralateral (non-injected) paw (B) for the first 30 min (P < 0.001). (C, D) The same as in A and B for von Frey mechanically evoked withdrawal threshold of the injected and non-injected paws. Note pretreatment with saline following intraplantar injection of CA resulted in thermal and mechanical hyperalgesia that persisted beyond 2 h (the brown dashed line). The thin black arrow indicates the time of injection of NSAIDs and the bold arrow indicates the time of injection of CA. BL – pre-injection baseline (reproduced from Tsagareli et al., 2018).

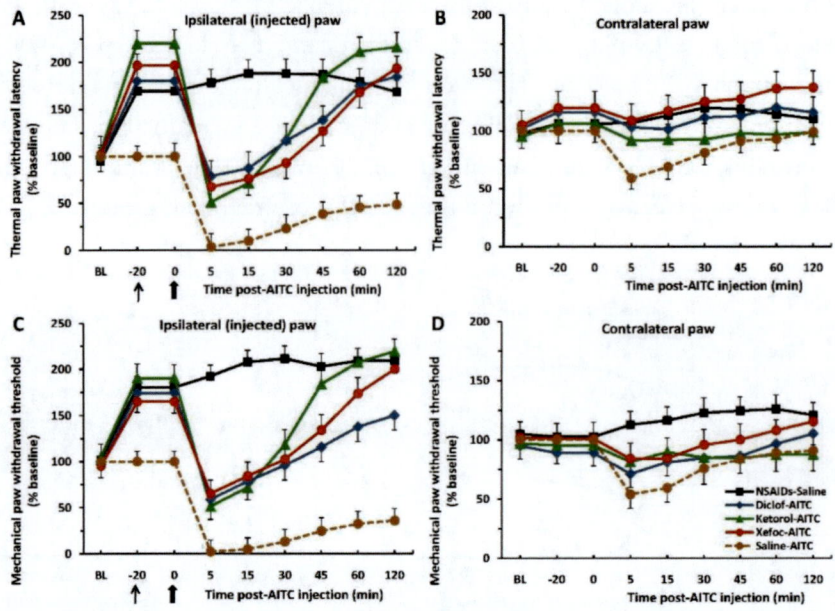

Figure 30. (A) Dynamics of the thermal paw withdrawal latency after NSAIDs pretreatment following ipsilateral intraplantar injection of AITC and vehicle. There are significant effects in AITC groups *vs.* vehicle control and contralateral (non-injected) paw (B) for the first 30 min (P < 0.001). (C, D) The same as in A and B for von Frey mechanically evoked withdrawal threshold of the injected and non-injected paws. Note pretreatment with saline following intraplantar injection of AITC resulted in thermal and mechanical hyperalgesia that persisted beyond 2 h (the brown dashed line). The thin black arrow indicates the time of injection of NSAIDs and the bold arrow indicates the time of injection of AITC. BL−pre-injection baseline (reproduced from Tsagareli et al., 2018).

There was a weaker mirror image hyperalgesic effect on the contralateral paw following CA, AITC and capsaicin injections preceded by saline in both the thermal and mechanical paw withdrawal tests (Figures 29B, D, 30B, D, 31B, D). In particular, the difference between the averaged NSAIDs-CA groups and the saline-CA group is significant at min 5 in both the thermal (t = 5.146, p < 0.01) and mechanical test (t = 6.324, p < 0.001) (Figure 29B, D). For the AITC trials, the difference is significant for Hargreaves test (t =4.336, p < 0.05), but not for von Frey test (t =

3.341, P > 0.05), respectively (Figure 30B, D). For the capsaicin session, the maximal difference between the NSAIDs-capsaicin groups occurred at min 30 in the thermal withdrawal test (t = 5.221, p < 0.01) and at min 15 in the mechanical withdrawal test (t = 4.224, p < 0.05), respectively (Figure 31B, D).

8.1.1. Lack of Antihyperalgesic Effect Following Post-Treatment with NSAIDs

In the fourth and fifth set of experiments, we tested if post-treatment with NSAIDs, diclofenac, ketorolac and xefocam, attenuated hyperalgesia produced by CA or AITC. Each of the 7 groups of rats received intraplantar injection of saline in one hindpaw to establish baseline responses. Three days later, rats were injected with CA or AITC in the same hindpaw, followed 20 min later by injection of one of the NSAIDs or saline at the same site. Control animals received an intraplantar injection of saline followed 20 min later by injection of one of the NSAIDs at the same site. Post-treatment with NSAIDs did not attenuate thermal or mechanical hyperalgesia induced by either CA (Figure 32A, C) or AITC (Figure 33A, C). Thermal and mechanical hyperalgesia continued out to 2 h post-NSAID injection without returning to the baseline level (Figures 32A, C, 33A, C). Each of three control groups treated with intraplantar saline followed by NSAIDs showed significant analgesia compare to groups receiving CA or AITC (Figures 32A, C, 33A, C) (p < 0.001). Here we also found small mirror-image hyperalgesic effects for the saline-CA and saline-AITC groups that are significantly different from the NSAIDs groups at min 5 for the thermal (CA: t = 3.72, P < 0.05; AITC: t = 3.685, p < 0.05) and mechanical test (CA: t = 3.69, p < 0.05; AITC: t = 4.165, p < 0.05) (Figures 32B, D, 33B, D).

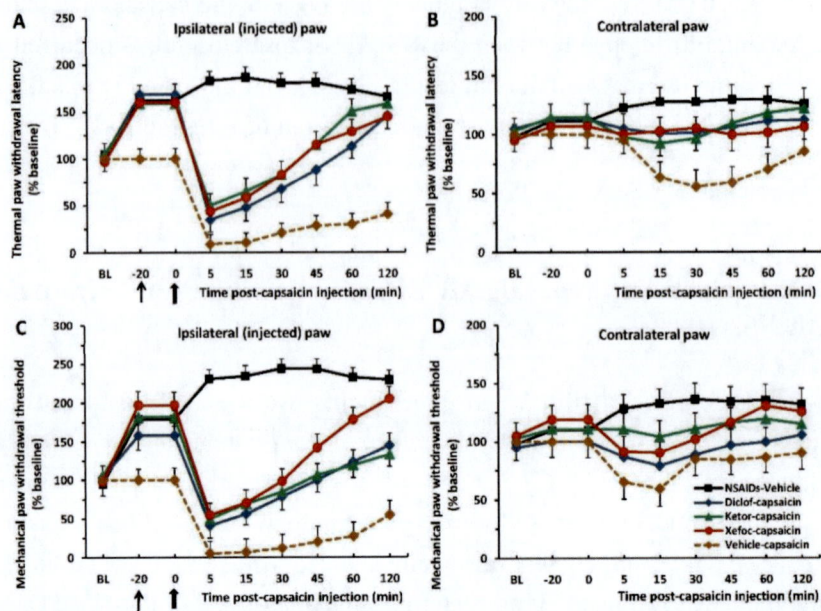

Figure 31. (A) Dynamics of the thermal paw withdrawal latency after NSAIDs pretreatment following ipsilateral intraplantar injection of capsaicin and vehicle. There are significant effects in capsaicin groups *vs.* vehicle control and contralateral (non-injected) paw (B) for the first 40 min (P < 0.001). (C, D) The same as in A and B for von Frey mechanically evoked withdrawal threshold of the injected and non-injected paws. Note pretreatment with saline following intraplantar injection of capsaicin resulted in thermal and mechanical hyperalgesia that persisted beyond 2 h (the brown dashed line). The thin black arrow indicates the time of injection of NSAIDs and a bold arrow indicates the time of injection of capsaicin. BL – pre-injection baseline (reproduced from Tsagareli et al., 2018).

It is well know that the therapeutic effects of NSAIDs are based on their inhibitory actions on cyclooxygenase enzymes (COX1 and COX2) and subsequent interference with metabolites of the arachidonic acid pathway. Previous reports have also described NSAIDs as blockers of TRP channels (Close et al., 2011). The latter are considered as targets for analgesic drug discovery. We found that NSAIDs pretreatment attenuates the duration (but not magnitude) of thermal and mechanical hyperalgesia induced by the TRPA1 channel agonists, CA and AITC and the TRPV1 channel agonist capsaicin, with a more rapid return to baseline in approximately 30 min.

Our data confirm previous reports. In particular, patch-clamp electrophysiology experiments showed that high doses of etodolac, an NSAID COX-2 inhibitor, could desensitize TRPA1 channels in heterologously expressing HEK293 cells and in the DRG neurons. Moreover, application of etodolac at clinically relevant drug plasma levels did not induce significant TRPA1 currents, but reduced the subsequent AITC-induced currents to 25% in HEK293 cells expressing TRPA1 (Wang et al., 2013). Earlier reports by Inoue et al. (2012) showed that pretreatment with etodolac inhibited Ca^{2+} influx induced by AITC in HEK-293 cells expressing mouse TRPA1 and in mouse DRG neurons. They also found that etodolac attenuates mechanical allodynia in a mouse model of neuropathic pain (Inoue et al., 2009; Ito et al., 2012).

In other experiments, it has recently shown that pyrazolone and pyrazolone derivatives (PDs, dipyrone, propyphenazone and antipyrine) selectively inhibited calcium responses and currents in TRPA1- expressing cells and acute nocifensive responses in mice evoked by reactive channel agonists (AITC, acrolein and H_2O_2). The two most largely used PDs, dipyrone and propyphenazone, attenuated TRPA1-mediated nociception and mechanical allodynia in models of inflammatory and neuropathic pain (formalin, carrageenan, partial sciatic nerve ligation) (Nassini et al., 2015). As the authors concluded, the PDs blockade of the nociceptive/ hyperalgesic effect produced by TRPA1 activation suggests that a similar pathway is attenuated by PDs in humans and that TRPA1 antagonists could be novel analgesics, devoid of the adverse hematological effects of PDs (Nassini et al., 2015).

However, post-treatment with NSAIDs, did not attenuate hyperalgesia produced by TRPA1agonists, CA and AITC; reduced thermal latencies and mechanical threshold had still not returned to baseline levels by 2 h. Thus, pre-treatment but not post-treatment with NSAIDs reduces the duration of thermal and mechanical hyperalgesia induced by TRP channel agonists.

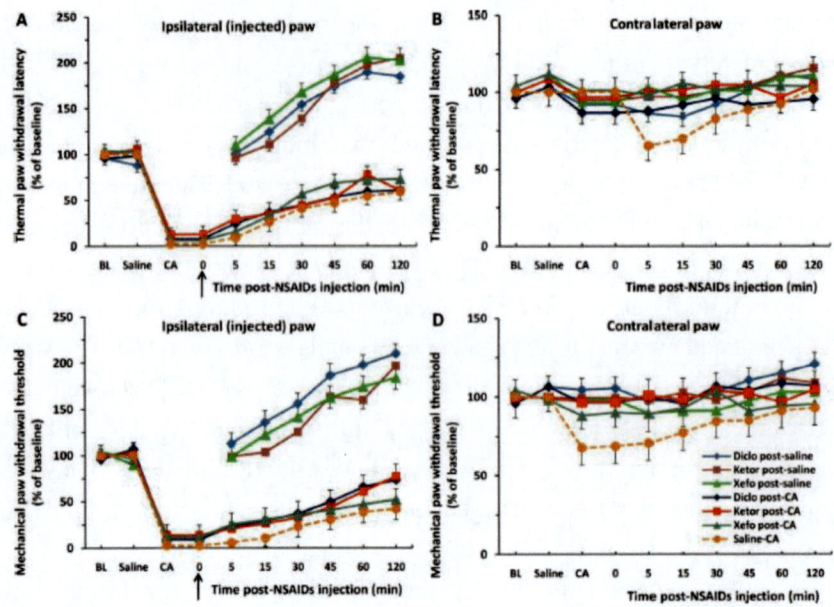

Figure 32. (A) Dynamics of the thermal paw withdrawal latency after NSAIDs or saline post-treatment. Preliminary experimental groups of rats ipsilaterally injected with CA. Saline control groups show strong analgesia compare to experimental groups ($P < 0.001$) for ipsilateral paw, but not for contralateral paw (B). (C, D) The same as in A and B for von Frey mechanically evoked withdrawal threshold of the injected and non-injected paws. The black arrow indicates the time of injection of NSAIDs. BL – pre-injection baseline. Note post-treatment with NSAIDs did not attenuate thermal or mechanical hyperalgesia induced by CA (reproduced from Tsagareli et al., 2018).

We speculate that NSAIDs may act at TRPV1 channels to shorten the time course of hyperalgesia induced by capsaicin that was presently observed. This is supported by previous studies reporting that NSAIDs inhibit TRPV1. Acetylsalicylic acid (ASA) inhibited capsaicin-induced currents in isolated DRG neurons (Kress et al., 1996), and inhibited the activation of small DRG neurons by noxious heat (Greffrath et al., 2002). A recent study reported that while ASA did not directly inhibit capsaicin-evoked calcium responses in TRPV1-expressing HEK293cells, it enhanced tachyphylaxis to repetitive capsaicin stimuli independent of COX inhibition (Maurer et al., 2014). The authors speculate that ASA may directly stabilize the desensitized state of TRPV1, or indirectly inhibit intracellular protein kinase signaling pathways (Maurer et al., 2014). Also, it was recently reported that NSAID-serotonin conjugates, such as ibuprofen-5-HT and flurbiprofen-5-HT, inhibit fatty acid amide hydrolase (a membrane-associated intracellular enzyme that degrades endocannabinoids), TRPV1, and COX2, while fenoprofen-5-HT and naproxen-5-HT showed activity as dual inhibitors of TRPV1 and COX2 (Rose et al., 2014).

These and the present data are consistent with a novel mechanism involving the anti-inflammatory and analgesic effects of NSAIDs that lead to a reduction in pain-like behavior following TRPA1 activation by CA or AITC, potentially through the inactivation of desensitization of the TRPA1 channel or through the reduction of prostaglandins, and could be used therapeutically for pain treatment (Nozadze et al., 2016a, 2016c, 2018; Tsagareli et al., 2018).

In collaboration with Italian colleagues from the laboratory of Pierangelo Geppetti, Florence University, we have just recently explored whether one of the widely used NSAIDs ibuprofen's reactive compound, ibuprofen-acyl glucuronide (IAG), contributes to the ibuprofen analgesic and anti- inflammatory actions by targeting TRPA1, using *in vitro* tools (TRPA1-expressing human and rodent cells) and *in vivo* mouse models of inflammatory pain (De Logu et al., 2019).

Ibuprofen, the first approved member of propionic acid derivatives, is a classical NSAID widely used for its analgesic and anti-inflammatory properties that is indicated to relieve inflammation and several types of pain, including headache, muscular pain, toothache, backache, and dysmenorrhea. Therapeutic effects of ibuprofen are attributed to inhibition of prostanoids synthesis by a non-selective, reversible inhibition of both COX1 and COX2. Shortly, we have found that IAG, but not ibuprofen, attenuates excitatory and pro-inflammatory responses in TRPA1-expressing cells *in vitro* and proalgesic responses *in vivo* elicited by reactive agonists of the channel. IAG also selectively attenuated the TRPA1-dependent component of the proalgesic responses evoked *in vivo* by formalin- or carrageenan-induced inflammation, thus underlying the hypothesis that TRPA1 targeting by IAG contributes to both analgesic and anti-inflammatory effects of ibuprofen.

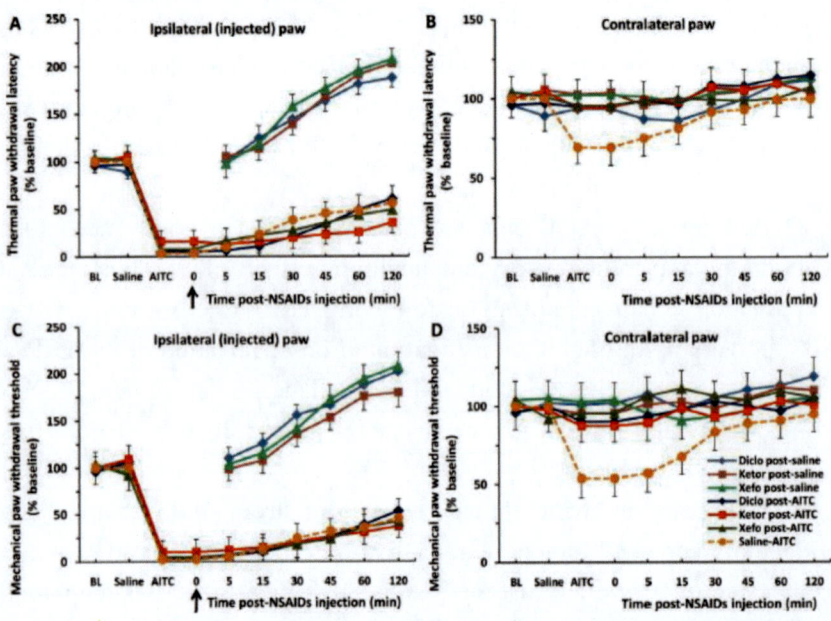

Figure 33. (A) Dynamics of the thermal paw withdrawal latency after NSAIDs or saline post-treatment. Preliminary experimental groups of rats ipsilaterally injected with AITC. Saline control groups show strong analgesia compare to experimental groups ($P < 0.001$) for ipsilateral paw, but not for contralateral paw (B). (C, D) The same as in A and B for von Frey mechanically evoked withdrawal threshold of the injected and non-injected paws. The black arrow indicates the time of injection of NSAIDs. BL – pre-injection baseline. Note post-treatment with NSAIDs did not attenuate thermal or mechanical hyperalgesia induced by AITC (reproduced from Tsagareli et al., 2018).

In particular, we found that IAG antagonizes the human recombinant TRPA1-HEK293 cells (Figure 34). Vehicle for IAG did not affect *per se* any calcium response in hTRPA1-HEK293 cells evoked by AITC (Figure 34B, left plot) however, IAG inhibited calcium response evoked by AITC (Figures 34B, right plot, 34I). Calcium responses to menthol (100 μM) were unaffected by IAG in 3C/K-Q hTRPA1-HEK293 cells (Figure 34E).

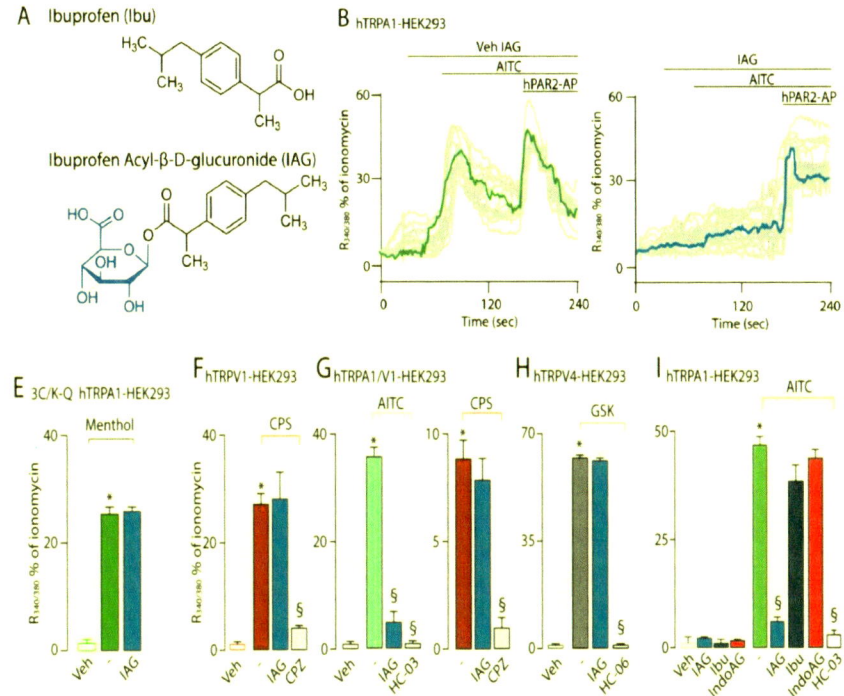

Figure 34. Ibuprofen-acyl-β-D-glucuronide (IAG) antagonizes the human recombinant TRPA1. (A) Chemical structure of ibuprofen (Ibu) and IAG. (B) Typical traces of the effect of IAG (100 μM) or its vehicle (Veh IAG) on calcium responses evoked by AITC (5 μM) in hTRPA1-HEK293. (E) Effect of IAG (100 μM) on the 3C/K-Q hTRPA1-HEK293 cells evoked by menthol (100 μM). (F) Effect of IAG (100 μM) and capsazepine (CPZ, 10 μM) on the calcium responses evoked by capsaicin (CPS, 0.1 μM) in hTRPV1-HEK293 cells. (G) Effect of IAG (100 μM), CPZ (10 μM) and HC-030031 (30 μM) on the calcium response evoked by CPS (0.1 μM) and AITC (10 μM) in hTRPA1/V1-HEK293 cells. (H) Effect of IAG (100 μM) and HC-067047 (HC-06, 10 μM) on the calcium response evoked by GSK1016790 A (GSK, 0.1 μM) in hTRPV4-HEK293 cells. (I) Effect of IAG (100 μM), Ibu (100 μM) and indomethacin acyl-β-D-glucuronide (IndoAG, 100 μM) on the calcium response evoked by AITC (5 μM) in hTRPA1-HEK293 cells. Values are mean ± s.e.m of n > 50 cells from at least 3 different experiments for each condition. Veh indicates vehicle of AITC, ACR, H_2O_2, icilin, $ZnCl_2$ and hPAR2-AP, dash (-) indicates vehicles of IAG, HC-03, ibu, CPZ, and HC-06. *$P < 0.05$ vs. Veh; §$P < 0.05$ vs. AITC, ACR, H_2O_2, icilin, $ZnCl_2$, CPS or GSK. One-way ANOVA and post-hoc Bonferroni's test (adapted from De Logu et al., 2019).

Like the selective TRPA1 antagonist, HC-030031, IAG inhibited in a concentration-dependent manner calcium response evoked by AITC [IC50, 30 (CI, 22–40) μM and 3 (CI, 1.4–6) μM, respectively] (data not shown). IAG reduced calcium responses evoked by additional reactive TRPA1 agonists, such as acrolein or hydrogen peroxide (H_2O_2), but did not affect the responses by non-reactive agonists, icilin and zinc chlorides ($ZnCl_2$), which do not act by binding key cysteine residues of TRPA1. HC-030031 abolished the calcium responses evoked by both reactive and non-reactive agonists. IAG did not attenuate the rapid calcium responses evoked by acute exposure to the activating peptide (AP) of the human PAR2 (hPAR2) (hPAR2-AP) (data not shown). This finding supports the selectivity of IAG. The ability of AIG to inhibit TRPA1 by binding key cysteine and lysine residues was further proved by the study of the mutated human TRPA1 (3C/K-Q hTRPA1), which lacks the cysteine and lysine residues, required for channel activation by reactive agonists, and which responds to menthol (Hinman et al., 2006; Macpherson et al., 2007; Trevisani et al., 2007).

Further selectivity of IAG for TRPA1 was robustly confirmed by a series of observations. In hTRPV1-HEK293, calcium responses to the TRPV1 agonist, capsaicin, were ablated by the TRPV1 selective antagonist, capsazepine, but were unaffected by IAG (Figure 34F). In hTRPA1/TRPV1-HEK293 co-expressing cells, responses to capsaicin were attenuated by capsazepine, but not by IAG, whereas responses to AITC were ablated by IAG and HC-03003, but not by capsazepine (Figure 34G). Moreover, in hTRPV4-HEK293 cells, calcium responses to the selective TRPV4 agonist, GSK1016790 A, were ablated by a TRPV4 antagonist, HC-067047, but were unaffected by IAG (Figure 34H). Ibuprofen did not evoke *per se* any calcium responses and did not affect the calcium responses evoked by AITC, acrolein or H_2O_2 in hTRPA1 HEK293 cells (Figure 34D, I). The glucuronidated metabolite of IAG, indomethacin neither evoked calcium response nor reduced the calcium response evoked by AITC in hTRPA1-HEK293 cells (Figure 34I).

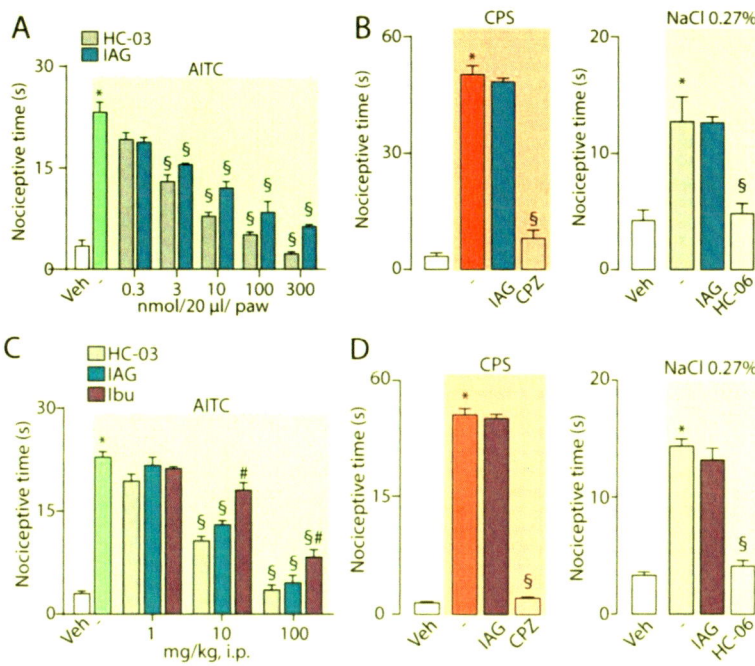

Figure 35. IAG inhibits nociceptive responses evoked by reactive TRPA1 agonists in mice. (A) Dose-dependent inhibitory effect of intraplantar (i.pl.) administration of IAG (20 μl/paw, 0.3–300 nmol) and HC-030031 (HC-03, 0.3 300 nmol) on the acute nociceptive response evoked by i.pl. AITC (20 nmol) in C57BL/6 J mice. (B) Effect of IAG (300 nmol), capsazepine (CPZ, 300 nmol) and HC-067047 (HC-06, 300 nmol) on the acute nociceptive response evoked by i.pl. CPS (1 nmol) and NaCl 0.27% in C57BL/6 J mice. (C) Dose-response inhibitory effect of intraperitoneal (i.p.) administration of IAG, Ibu and HC-03 (all, 1 – 100 mg/kg) on the acute nociceptive response evoked by i.pl. AITC (20 nmol) in C57BL/6 J mice. (D) Effect of i.p. IAG (100 mg/kg) CPZ (4 mg/kg) and HC-06 (10 mg/kg) on the acute nociceptive response evoked by i.pl. CPS (1 nmol) and NaCl 0.27% in C57BL/6 J mice. Values are mean ± s.e.m of n = 6 mice for each experimental condition. Veh indicates vehicle of CPS, NaCl 0.27%, ACR, $ZnCl_2$ and AITC, dash (-) indicates vehicles of IAG, HC- 03, ibu, CPZ and HC-06. *$P < 0.05$ vs. Veh; §$P < 0.05$ vs. CPS or NaCl 0.27%, ACR and $ZnCl_2$, #$P < 0.05$ vs. HC-03 and IAG. One way ANOVA and post-hoc Bonferroni's test (adapted from De Logu et al., 2019).

In vivo behavior experiments showed that the intraplantar (i.pl.) (20 µl/paw) administration of IAG or HC-030031 dose-dependently reduced [ID50 of 4 (CI, 2–9) nmol, and ID50, 8 (CI, 3–23) nmol, respectively] the acute nociceptive response evoked by the injection of AITC (i.pl.). Maximum inhibition on the nociceptive responses evoked by AITC (i.pl.) was 72% ± 2% for IAG and, 89% ± 1.7% for HC-030031 (n = 6, p < 0.05) (Figure 35A). Acute nociceptive responses induced by intraplantar capsaicin and hypotonic solution (TRPV1 and TRPV4-mediated responses, respectively) were attenuated by injection of the respective channel antagonists, capsazepine and HC-067047, but were unaffected by IAG (all i.pl.) (Figure 35B).

The systemic (intraperitoneal, i.p.) administration of HC-030031, IAG and ibuprofen dose-dependently [ID50s 7 (CI, 4–14) mg/kg, 10 (CI, 4–20) mg/kg and ID50s 27 (CI, 8–90) mg/kg, respectively] reduced the nociceptive responses to AITC (i.pl.) (Figure 35C). Maximum inhibition by ibuprofen (42% ± 3%) was lower than those produced by IAG (76% ± 4%) and HC-030031 (83 ± 4%) (all 100 mg/kg, n = 6 each, P < 0.05 ibuprofen *vs.* both IAG and HC-030031) (Figure 35C).

Finally, systemic (i.p.) IAG did not affect the nociceptive responses evoked by either capsaicin or a hypotonic solution, which, however, were attenuated by the TRPV1 and TRPV4 antagonists, capsazepine and HC067047, respectively (Figure 35D).

As stated above, the COXs inhibitor ibuprofen is widely used as a first line treatment for the relief of pain and inflammation. Glucuronide metabolites, including those generated from ibuprofen, are generally considered inactive and rapidly excreted compounds. However, acyl glucuronides, undergoing hydrolysis, acyl migration and molecular rearrangement, exhibit chemical reactivity that allows them to covalently bind various macromolecules, among them TRPA1 channels. A major finding of this study is that IAG, but not its parent compound, ibuprofen, antagonizes the pro-algesic TRPA1. This conclusion derives, primarily, from the *in vitro* pharmacological profile of IAG, which, unlike ibuprofen, selectively inhibits the recombinant and native human TRPA1 and the native rodent channel in nociceptors. Failure of the acyl derivative of indomethacin to affect channel activity underlines the unique ability of IAG to target TRPA1.

Additional *in vivo* data strengthen the conclusion obtained from our *in vitro* findings. Local injection of IAG in the mouse hind paw prevented acute nociception elicited by local administration of the reactive TRPA1 agonists, AITC and acrolein, but was ineffective against TRPV1 or TRPV4 agonists, indicating selectivity. Notably, local injection of ibuprofen in the mouse hind paw did not affect AITC or acrolein-evoked nociception. It is possible that following i.pl. ibuprofen no IAG is generated locally and the action of TRPA1 agonists remains unopposed. However, about 10 – 15% of systemic ibuprofen is converted into IAG. Thus, liver metabolism of a high dose of ibuprofen may produce IAG levels such as to guarantee a local concentration sufficient for inhibiting TRPA1. This hypothesis is supported by the observation that a high dose of systemic ibuprofen produced a partial attenuation of the nociception evoked by AITC.

In summary, presented findings add new insights into the antinociceptive/anti-hyperalgesic and anti-inflammatory activity of ibuprofen which, in addition to COX inhibition, attenuates TRPA1 activity *via* IAG generation. This novel mechanism of ibuprofen/IAG indirectly underlines the TRPA1 contribution to acute nociception and delayed allodynia in various models of inflammatory pain. Further studies are needed to establish whether TRPA1 antagonism by IAG contributes to the therapeutic effect of ibuprofen in pain and inflammation in humans, and whether IAG may have an efficacy and safety profile different from its parent drug (De Logu et al., 2019).

Chapter 9

CONCLUDING REMARKS AND PERSPECTIVES

The discovery of the *trp* gene superfamily almost twenty five years ago and which encodes the TRP cation channels has opened a highly exciting field to understand physiological and pathophysiological basic mechanisms of cell functions, which were until this time inaccessible. TRP channels comprise one of the most rapid growing research topics in related to ion channels including channelopathies and translational medicine (Nilius, Flockerzi, 2014).

Recent advancement in the field of cryo-EM has enabled near atomic resolution structures of some of the TRP channels, giving a glimpse into their molecular architecture. The structural information will help in establishing some of the functional characteristics of these channels and will also be beneficial for drug development. Resolving more structures will be important for the better understanding this super-family of ion channels in physiological processes (Samanta et al., 2018).

Eleven mutant TRP channels cause a spectrum of 16 human diseases, additionally emphasizing their essential role *in vivo*. Moreover, TRP channels are important pharmacological targets for specific novel therapeutic treatment options for patients. Along these lines, specific TRP modulators have been identified in recent years and are now tested *in vitro* and *in vivo* against symptoms caused by dysfunctional TRP proteins or pathophysiological processes such as pain, chronic inflammation, fibrosis, and edema, which occur if normal physiological responses are out of control (Dietrich, 2019b; Moran, 2018). Understanding of the mechanistic background of these channel malfunctions not only is essential for a treatment of these diseases but may also help to understand the functioning of these exciting proteins and their involvement in different signaling cascades.

The enthusiasm that TRP channels may solve many of health problems should be qualified by a more rigorous investigation of TRP channel functions in more refined and suitable models. At the same time, obviously, we have to admit that in many diseases the role of TRPs might be somewhat over-interpreted/overstressed. Better pathophysiological models which consider the difficulties to handle TRP must be developed. We are still in the very beginning of understanding disease-causing dysfunctions of TRP channels (Nilius, Flockerzi, 2014).

Because of all the contradictions and gaps in our understanding of the TRP, we have to admit that the exciting field of investigation is open, but the current state of research on the TRP channel is still at the beginning.

REFERENCES

Abe, J., Hosokawa, H., Sawada, Y., Matsumura, K. & Kobayashi, S. (2006). Ca^{2+}-dependent PKC activation mediates menthol-induced desensitization of transient receptor potential M8. *Neurosci. Letters*, ***397***(1-2), 140–144 (doi: 10.1016/j.neulet.2005.12.005).

Albin, K. C., Carstens, M. I. & Carstens, E. (2008). Modulation of oral heat and cold pain by irritant chemicals. *Chem. Senses*, ***33***(1), 3-15.

Almaraz, L., Manenschijn, J. A., de la Peña, E. & Viana, F. (2014). TRPM8. In: B. Nilius, V. Flockerzi (eds.) *Mammalian Transient Receptor Potential (TRP) Cation Channels.* Handb. Exp. Pharmacol., Springer, ***222***, 547-579 (doi: 10.1007/978-3-642-54215-2_22).

Alonso-Carbajo, L., Alpizar, Y. A., Startek, J. B., López-López, J. R., Pérez-García, M. T. & Talavera, K. (2019). Activation of the cation channel TRPM3 in perivascular nerves induces vasodilation of resistance arteries. *J. Mol. Cell. Cardiol.*, ***129***, 219-230 (doi: 10.1016/j.yjmcc.2019.03.003).

Al Alawi, A. M., Majoni, S. W. & Falhammar, H. (2018). Magnesium and human Health: Perspectives and research directions. *Int. J. Endocrinol.*, art. 9041694 (doi: 10.1155/2018/9041694).

Bandell, M., Story, G. M., Hwang, S. W., Viswanath, V., Eid, S. R., Petrus, M. J., Earley, T. J. & Patapoutian, A. (2004). Noxious cold ion channel TRPA1 is activated by pungent compounds and bradykinin. *Neuron*, ***41***(6), 849–857.

Bartfai, T. (2016). Cool by neuronal decision. *Science*, *353*(6306), 1363–1364 (doi: 10.1126/science.aai8465).

Basbaum, A. I., Bautista, D. M., Scherrer, G. & Julius, D. (2009) Cellular and molecular mechanisms of pain. *Cell*, *139*(2), 267-284 (doi: 10.1016/j.cell.2009.09.028).

Bautista, D. M., Jordt, S. E., Nikai, T., Tsuruda, P. R., Read, A. J., Poblete, J., Yamoah, E. N., Basbaum, A. I. & Julius, D. (2006). TRPA1 mediates the inflammatory actions of environmental irritants and proalgesic agents. *Cell*, *124* (6), 1269-1282 (doi: 10.1016/j.cell. 2006. 02.023).

Bautista, D. M., Siemens, J., Glazer, J. M., Tsuruda, P. R., Basbaum, A. I., Stucky, C. L., Jordt, S. E. & Julius, D. (2007). The menthol receptor TRPM8 is the principal detector of environmental cold. *Nature*, *448*(7150), 204–208 (doi: 10. 1038/nature05910).

Belmonte, C., Brock, J. & Viana, F. (2009). Converting cold into pain. *Exp. Brain Res.*, *196*(1), 13–30 (doi: 10.1007/s00221-009-1797-2).

Belmonte, C. & Viana, F. (2008). Molecular and cellular limits to somatosensory specificity. *Mol. Pain*, *4*, art. 14 (doi: 10.1186/1744-8069-4-14).

Benemei, S. & Dussor, G. (2019). TRP channels and migraine: Recent developments and new therapeutic opportunities. *Pharmaceuticals* (Basel), *12*(2), E54 (doi: 10.3390/ph12020054).

Benemei, S., Patacchini, R., Trevisani, M. & Geppetti, P. (2015). TRP channels. *Curr. Opin. Pharmacol.*, *22*, 18-23 (doi: 10.1016/ j.coph. 2015.02.006).

Bevan, S., Quallo, T. & Andersson, D. A. (2014). TRPV1. In: B. Nilius, V. Flockerzi (eds.) *Mammalian Transient Receptor Potential (TRP) Cation Channels, Handb. Exp. Pharmacol.*, Springer, *222*, 207-245 (doi: 10.1007/978-3-642-54215-2_9).

Birnbaumer, L. (2009). The TRPC class of ion channels: A critical review of their roles in slow, sustained increases in intracellular Ca^{2+} concentrations. *Annu. Rev. Pharmacol. Toxicol.*, *49*, 395–426 (doi: 10. 1146/annurev.pharmtox.48.113006.094928).

Blanquart, S., Borowiec, A. S., Delcourt, P., Figeac, M., Emerling, C. A., Meseguer, A. S., Roudbaraki, M., Prevarskaya, N. & Bidaux, G. (2019). Evolution of the human cold/menthol receptor, TRPM8. *Mol. Phylogenet. Evol.*, *136*, 104-118 (doi: 10.1016/j.ympev.2019.04.01).

Bollimuntha, S., Selvaraj, S. & Singh, B. B. (2011). Emerging roles of canonical TRP channels in neuronal function. *Adv. Exp. Med. Biol.*, *704*, 573–593 (doi: 10.1007/978-94-007-0265-3_31).

Boulant, J. A. (2000). Role of the preoptic-anterior hypothalamus in thermoregulation and fever. *Clin. Infect. Dis.*, *31*(Suppl. 5), S157–S161.

Bouron, A., Chauvet, S., Dryer, S. & Rosado, J. A. (2016). Second messenger-operated calcium entry through TRPC6. In: J.A. Rosado (ed.) *Calcium Entry Pathways in Non-excitable Cells*, Chapter 10. Adv. Exp. Med. Biol., *898*, 201-249 (doi: 10.1007/978-3-319-26974-0_10_).

Broad, L. M., Mogg, A. J., Eberle, E., Tolley, M., Li, D. L. & Knopp, K. L. (2016). TRPV3 in drug development. *Pharmaceuticals* (Basel), *9*(3), E55 (doi: 10.3390/ph9030055).

Bráz, J. M. & Basbaum, A. I. (2010). Differential ATF3 expression in dorsal root ganglion neurons reveals the profile of primary afferents engaged by diverse noxious chemical stimuli. *Pain*, *150*(2), 290–301 (doi: 10.1016/j.pain.2010.05.005).

Carstens, E., Albin, K. C., Simons, C. T. & Iodi Carstens, M. (2007). Time course of self-desensitization of oral irritation by nicotine and capsaicin. *Chem. Senses*, *32*(9), 811–816 (doi: 10.1093/chemse/bjm048).

Carstens, E., Klein, A., Iodi Carstens, M., Tsagareli, M. G., Tsiklauri, N., Gurtskaia, G. & Abzianidze, E. (2010). Effects of thermosensitive transient receptor potential (TRP) channel agonists on the neural coding of touch, temperature and pain sensation. *Proc. 9th "Gagra Talks". Int. Confer. Fundam. Probl. Neurosci.* Tbilisi, pp. 267-280.

Caterina, M. J. (2007). Transient receptor potential ion channels as participants in thermosensation and thermoregulation. *Am. J. Physiol.*

Regul. Integr. Comp. Physiol., *292*, R64–R76 (doi: 10.1152/ajpregu. 0046.2006).

Caterina, M. J., Leffler, A., Malmberg, A. B., Martin, W. J., Trafton, J., Petersen-Zeitz, K. R., Koltzenburg, M., Basbaum, A. I. & Julius, D. (2000). Impaired nociception and pain sensation in mice lacking the capsaicin receptor. *Science*, *288*(5464), 306–313 (doi: 10.1126/science. 288.5464.306).

Caterina, M. J., Schumacher, M. A., Tominaga, M., Rosen, T. A., Levine, J. D. & Julius, D. (1997). The capsaicin receptor: a heat-activated ion channel in the pain pathway. *Nature*, *389*(6653), 816-824 (doi: 10. 1038/39807).

Chen, C. C., Butz, E. S., Chao, Y. K., Grishchuk, Y., Becker, L., Heller, S., Slaugenhaupt, S. A., Biel, M., Wahl-Schott, C. & Grimm, C. (2017). Small molecules for early endosome-specific patch clamping. *Cell Chem. Biol.*, *24*(7), 907-916 (doi: 10.1016/j.chembiol.2017.05.025).

Chen, Y., Zhang, X., Yang, T., Bi, R., Huang, Z., Ding, H., Li, J. & Zhang, J. (2019). Emerging structural biology of TRPM subfamily channels. *Cell Calcium.*, *79*, 75-79 (doi: 10.1016/j.ceca.2019.02.011).

Christensen, A. P. & Corey, D. P. (2007). TRP channels in mechanosensation: direct or indirect activation? *Nature Rev. Neurosci.*, *8*(7), 510-521 (doi: 10.1038/nrn2149).

Chuang, H. H., Prescott, E. D., Kong, H., Shields, S., Jordt, S. E., Basbaum, A. I., Chao, M. V. & Julius, D. (2001). Bradykinin and nerve growth factor release the capsaicin receptor from PtdIns (4,5) P2-mediated inhibition. *Nature*, *411*(6840), 957–962 (doi: 10.1038/ 35082088).

Chubanov, V., Ferioli, S. & Gudermann, T. (2017). Assessment of TRPM7 functions by drug-like small molecules. *Cell Calcium*, *67*, 166-173 (doi: 10.1016/j.ceca.2017.03.004).

Chubanov, V. & Guderman, T. (2014). TRPM6. In: B. Nilius, V. Flockerzi (eds.) *Mammalian Transient Receptor Potential (TRP) Cation Channels*. Handb. Exp. Pharmacol., Springer, *222*, 503-520 (doi: 10. 1007/978-3-642-54215-2_20).

Clapham, D. E. (2003). TRP channels as cellular sensors. *Nature*, **426**(6966), 517-524 (doi: 10.1038/nature02196).

Clapham, D. E. (2015). Pain-sensing TRPA1 channel resolved. *Nature*, **520**(7548), 439-41 (doi: 10.1038/nature 14383).

Cliff, M. A. & Green, B. G. (1994). Sensory irritation and coolness produced by menthol: Evidence for selective desensitization of irritation. *Physiol. Behav.*, **56**(5), 1021–1029.

Cliff, M. A. & Green, B. G. (1996). Sensitization and desensitization to capsaicin and menthol in the oral cavity: interactions and individual differences. *Physiol. Behav.*, **59**(3), 487–494.

Close, C., Straub, I., Riehle, M., Ranta, F., Krautwurst, D., Ullrich, S., Meyerhof, W. & Harteneck, C. (2011). Fenamates as TRP channel blockers: mefenamic acid selectively blocks TRPM3. *Brit. J. Pharmacol.*, **162**(8), 1757–176 (doi: 10.1111/j.1476-5381.2010.01186.x).

Colbert, H. A., Smith, T. L. & Bargmann, C. I. (1997). OSM-9, a novel protein with structural similarity to channels, is required for olfaction, mechanosensation, and olfactory adaptation in *Caenorhabditis elegans. J. Neurosci.*, **17** (21), 8259-8269.

Colburn, R. W., Lubin, M. L., Stone, Jr. D. J., Wang, Y., Lawrence, D., D'Andrea, M. R., Brandt, M. R., Liu, Y., Flores, C. M. & Qin, N. (2007). Attenuated cold sensitivity in TRPM8 null mice. *Neuron*, **54**(3), 379–386 (doi: 10.1016/j.neuron.2007.04.017).

Conti, B. (2018). Molecular basis of central thermosensation. In: A.A. Romanovsky (ed.), Thermoregulation: From Basic Neuroscience to Clinical Neurology, chapter 8, Part I. *Handb. Clin. Neurol.*, **156** (3rd series), 129-133 (doi: 10.1016/B978-0-444-63912-7.00008-4).

Corey, D. P., Garcia-Anoveros, J., Holt, J. R., Kwan, K. Y., Lin, S. Y., Vollrath, M. A., Amalfitano, A., Cheung, E. L. M., Derfler, B. H., Duggan, A., Geleoc, G. S. G., Gray, P. A., Hoffman, M. P., Rehm, H. L., Tamasauskas, D. & Zhang, D. S. (2004). TRPA1 is a candidate for the mechanosensitive transduction channel of vertebrate hair cells. *Nature*, **432**(7018), 723–730 (doi: 10.1038/nature03066).

Craig, A. D. & Dostrovsky, J. O. (2001). Differential projections of thermoreceptive and nociceptive lamina I trigemino-thalamic and spinothalamic neurons in the cat. *J. Neurophysiol.*, *86*(2), 856–870 (doi: 10.1152/jn.2001.86.2.856).

Cuajungco, M. P., Silva, J., Habibi, A. & Valadez, J. A. (2016). The mucolipin-2 (TRPML2) ion channel: a tissue-specific protein crucial to normal cell function. *Pflugers Arch.*, *468*(2), 177-192 (doi: 10.1007/s00424-015-1732-2).

Dai, Y. (2016). TRPs and pain. *Semin. Immunopathol.*, *38*(3), 277-291 (doi: 10.1007/s00281-015-0526-0).

Danhof, M., Klein, K., Stolk, P., Aitken, M. & Leufkens, H. (2018). The future of drug development: the paradigm shift towards systems therapeutics. *Drug Discov. Today*, *23*(12), 1990-1995 (doi: 10.1016/j.drudis.2018. 09. 002).

Darby, W. G., Grace, M. S., Baratchi, S. & McIntyre, P. (2016). Modulation of TRPV4 by diverse mechanisms. *Int. J. Biochem. Cell Biol.*, *78*, 217-228 (doi: 10.1016/j.biocel.2016.07.012).

Davis, J. B., Gray, J., Gunthorpe, M. J., Hatcher, J. P., Davey, P. T., Overend, P., Harries, M. H., Latcham, J., Clapham, C., Atkinson, K., Hughes, S. A., Rance, K., Grau, E., Harper, A. J., Pugh, P. L., Rogers, D. C., Bingham, S., Randall, A. & Sheardown, S. A. (2000). Vanilloid receptor-1 is essential for inflammatory thermal hyperalgesia. *Nature*, *405* (6783), 183–187 (doi: 10.1038/35012076).

de Andrade, E. L., Meotti, F. C. & Calixto, J. B. (2012). TRPA1 antagonists as potential analgesic drugs. *Pharmacol. Therap.*, *133*(2), 189-204 (doi: 10.1016/j.pharmthera.2011.10.008).

De Logu, F., Li Puma, S., Landini, L., Tuccinardi, T., Poli, G., Preti, D., DeSiena, G., Patacchini, R., Tsagareli, M. G., Geppetti, P. & Nassini, R. (2019). The acyl-glucuronide metabolite of ibuprofen has analgesic and anti-inflammatory effects via the TRPA1 channel. *Pharmacol. Res.*, *142*, 127-139 (doi: 10.1016/j.phrs.2019.02.019).

De Logu, F., Patacchini, R., Fontana, G. & Geppetti, P. (2016). TRP functions in the broncho-pulmonary system. *Semin. Immunopathol.*, *38*(3), 321-329 (doi: 10.1007/s00281-016-0557-1).

Dessirier, J. M., O'Mahony, M. & Carstens, E. (2001). Oral irritant properties of menthol: Sensitizing and desensitizing effects of repeated application and cross-desensitization to nicotine. *Physiol. Behav.*, *73*(1-2), 25–36 (doi: 10. 016/s0031-9384(01)00431-0).

Dhaka, A., Earley, T. J., Watson, J. & Patapoutian, A. (2008). Visualizing cold spots: TRPM8-expressing sensory neurons and their projections. *J. Neurosci.*, *28*(3), 566–575 (doi: 10.1523/JNEUROSCI.3976-07. 2008).

Dhaka, A., Murray, A. N., Mathur, J., Earley, T. J., Petrus, M. J. & Patapoutian, A. (2007). TRPM8 is required for cold sensation in mice. *Neuron*, *54*(3), 371–378 (doi: 10.1016/j.neuron.2007.02.024).

Dhaka, A., Viswanath, V. & Patapoutian, A. (2006). TRP ion channels and temperature sensation. *Annu. Rev. Neurosci.*, *29*, 135-161 (doi: 10. 1146/annurev.neuro.29.051605.112958).

Dietrich, A. (2019a). Modulators of transient receptor potential (TRP) channels as therapeutic options in lung disease. *Pharmaceuticals*, *12*(1), pii: E23 (doi: 10.3390/ph12010023).

Dietrich, A. (2019b). Transient receptor potential (TRP) channels in health and disease. *Cells*, *8*(5), pii: E413 (doi: 10.3390/cells8050413).

Dietrich, A., Fahlbusch, M. & Gudermann, T. (2014). Classical transient receptor potential 1 (TRPC1): Channel or channel regulator? *Cells*, *3*(4), 939-962 (doi: 10.3390/cells3040939).

Dietrich, A. & Gudermann, T. (2014). TRPC6: Physiological function and pathophysiological relevance. In: B. Nilius, V. Flockerzi (eds.) *Mammalian Transient Receptor Potential (TRP) Cation Channels.* Handb. Exp. Pharmacol., Springer, *222*, 157-188 (doi: 10.1007/978-3-642-54215-2_7).

Dietrich, A., Steinritz, D. & Gudermann, T. (2017). Transient receptor potential (TRP) channels as molecular targets in lung toxicology and associated diseases. *Cell Calcium.*, *67*, 123-137 (doi: 10.1016/j.ceca. 2017.04.005).

Duan, J., Li, Z., Li, J., Hulse, R. E., Santa-Cruz, A., Valinsky, W. C., Abiria, S. A., Krapivinsky, G., Zhang, J. & Clapham, D. E. (2018a). Structure of the mammalian TRPM7, a magnesium channel required during embryonic development. *Proc. Natl. Acad. Sci. USA.*, *115*(35), E8201-E8210 (doi: 10.1073/pnas.1810719115).

Duan, J., Li, Z., Li, J., Santa-Cruz, A., Sanchez-Martinez, S., Zhang, J. & Clapham, D. E. (2018b). Structure of full-length human TRPM4. *Proc. Natl. Acad. Sci. USA.*, *115*(10), 2377-2382 (doi: 10.1073/pnas.1722038115).

Duan, J., Li, J., Zeng, B., Chen, G. L., Peng, X., Zhang, Y., Wang, J., Clapham, D. E., Li, Z. & Zhang, J. (2018c). Structure of the mouse TRPC4 ion channel. *Nature Commun.*, *9*(1), 3102 (doi: 10.1038/s41467-018-05247-9).

De Felice, M., Eyde, N., Dodick, D., Dussor, G. O., Ossipov, M. H., Fields, H. L. & Porreca, F. (2013). Capturing the aversive state of cephalic pain preclinically. *Ann. Neurol.*, *74*(2), 257–265 (doi: 10.1002/ana.23922).

Di Paola, S., Scotto-Rosato, A. & Medina, D. L. (2018). TRPML1: The $Ca^{(2+)}$retaker of the lysosome. *Cell Calcium.*, *69*, 112-121 (doi: 10.1016/j.ceca.2017.06.006).

Eid, S. R., Crown, E. D., Moore, E. L., Liang, H. A., Choong, K. C., Dima, S., Henze, D. A., Kane, S. A. & Urban, M. O. (2008). HC-030031, A TRPA1 selective antagonist, attenuates inflammatory- and neuropathy-induced mechanical hypersensitivity. *Mol. Pain*, *4*, art. 48 (doi: 10.1186/1744-8069-4-48).

Faouzi, M. & Penner, R. (2014). TRPM2. In: B. Nilius, V. Flockerzi (eds.) *Mammalian Transient Receptor Potential (TRP) Cation Channels.* Handb. Exp. Pharmacol., Springer, *222*, 403-426 (doi: 10.1007/978-3-642-54215-2_16).

Farco, J. A. & Grundmann, O. (2013). Menthol – pharmacology of an important naturally medicinal "cool". *Mini Rev. Med. Chem.*, *13*(1), 124-131.

Fecher-Trost, C., Weissgerber, P. & Wissenbach, U. (2014). TRPV6 Channels. In: B. Nilius, V. Flockerzi (eds.) Mammalian Transient Receptor Potential (TRP) Cation Channels. *Handb. Exp. Pharmacol.*, Springer, *222*, 359-384 (doi: 10.1007/978-3-642-54215-2_14).

Fecher-Trost, C., Wissenbach, U. & Weissgerber, P. (2017). TRPV6: From identification to function. *Cell Calcium*, *67*, 116-122 (doi: 10.1016/j.ceca.2017.04.006).

Fischer, M. J., Balasuriya, D., Jeggle, P., Goetze, T. A., McNaughton, P. A., Reeh, P. W. & Edwardson, J. M. (2014). Direct evidence for functional TRPV1/TRPA1 heteromers. *Pflugers Arch.*, *466*(12), 2229–2241 (doi: 10.1007/s00424-014-1497-z).

Fleig, A. & Chubanov, V. (2014). TRPM7. In: B. Nilius, V. Flockerzi (eds.) Mammalian Transient Receptor Potential (TRP) Cation Channels. *Handb. Exp. Pharmacol.*, Springer, *222*, 521-546 (doi: 10.1007/978-3-642-54215-2_21).

Flockerzi, V. & Nilius, B. (2014). TRPs: Truly remarkable proteins. In: B. Nilius, V. Flockerzi (eds.) Mammalian Transient Receptor Potential (TRP) Cation Channels, *Handb. Exp. Pharmacol.*, Springer, *222*, 1-12 (doi: 10.1007/978-3-642-54215-2_1).

Fowler, M. A. & Montell, C. (2013). Drosophila TRP channels and animal behavior. *Life Sci.*, *92*(8-9), 394-403 (doi: 10.1016/j.lfs.2012.07.029).

Freichel, M. & Tsvilovskyy, Camacho-Londoño J. E. (2014). TRPC4- and TRPC4-containing channels. In: B. Nilius, V. Flockerzi (eds.) *Mammalian Transient Receptor Potential (TRP) Cation Channels.* Handb. Exp. Pharmacol., Springer, *222*, 85-128 (doi: 10.1007/978-3-642-54215-2_5).

Galeotti, N., Di Cesare Mannelli, L., Mazzanti, G., Bartolini, A. & Ghelardini, C. (2002). Menthol: a natural analgesic compound. *Neurosci. Letters*, *322*(3), 145–148 (doi: 10.1016/s0304-3940(01)02527-7).

Galeotti, N., Ghelardini, C., Di Cesare Mannelli, L., Mazzanti, G., Baghiroli, L. & Bartolini, A. (2001). Local anesthetic activity of (+) and (−)-menthol. *Planta Med.*, *67*(2), 174–176 (doi: 10.1055/s-2001-11515).

Galindo, T., Reyna, J. & Weyer, A. (2018). Evidence for transient receptor potential (TRP) channel contribution to arthritis pain and pathogenesis. *Pharmaceuticals* (Basel), *11*(4), pii: E105 (doi: 10.3390/ph11040105).

Gao, Y. & Liao, P. (2019). TRPM4 channel and cancer. *Cancer Letters*, *454*, 66-69 (doi: 10.1016/j.canlet.2019.04.012).

García-Añoveros, J. & Wiwatpanit, T. (2014). TRPML2 and mucolipin evolution. In: B. Nilius, V. Flockerzi (eds.) *Mammalian Transient Receptor Potential (TRP) Cation Channels.* Handb. Exp. Pharmacol., *222*: 647-658 (doi: 10.1007/978-3-642-54215-2_25).

Garcia-Elias, A., Mrkonjić, S., Jung, C., Pardo-Pastor, C., Vicente, R. & Valverde, M. A. (2014). The TRPV4 channel. In: B. Nilius, V. Flockerzi (eds.) *Mammalian Transient Receptor Potential (TRP) Cation Channels*, Handb. Exp. Pharmacol., Springer, *222*, 293-319 (doi: 10.1007/978-3-642-54215-2_12).

Gees, M., Colsoul, B. & Nilius, B. (2012). The role of transient receptor potential cation channels in Ca^{2+} signaling. *Cold Spring Harb. Perspect. Biol.*, *2*, a003962 (doi: 10.1101/cshperspect.a003962).

Geppetti, P., Benemei, S. & De Cesaris, F. (2015). CGRP receptors and TRP channels in migraine. *J. Headache Pain*, *16*, (Suppl 1), A21 (doi: 10.1186/1129-2377-16-S1-A21).

Gilchrist, H. D., Allard, B. L. & Simone, D. A. (1996). Enhanced withdrawal responses to heat and mechanical stimuli following intraplantar injection of capsaicin in rats. *Pain*, *67*(1), 179–188.

González-Muñiz, R., Bonache, M. A., Martín-Escura, C. & Gómez-Monterrey, I. (2019). Recent progress in TRPM8 modulation: An update. *Int. J. Mol. Sci.*, *20*(11), pii: E2618 (doi: 10.3390/ijms20112618).

Goretzki, B., Glogowski, N. A., Diehl, E., Duchardt-Ferner, E., Hacker, C., Gaudet, R. & Hellmich, U. A. (2018). Structural basis of TRPV4 N-terminus interaction with syndapin/PACSIN1-3 and PIP_2. *Structure*, *26*(12), 1583-1593, e5 (doi: 10.1016/j.str.2018.08.002).

Grace, M. S., Bonvini, S. J., Belvisi, M. G. & McIntyre, P. (2017). Modulation of the TRPV4 ion channel as a therapeutic target for disease. *Pharmacol. Ther.*, *177*, 9-22 (doi: 10.1016/j.pharmthera.2017.02.019).

Grayson, T. H., Murphy, T. V. & Sandow, S. L. (2017). Transient receptor potential canonical type 3 channels: Interactions, role and relevance –

A vascular focus. *Pharmacol. Ther.*, **74**, 79-96 (doi: 10.1016/j.pharmthera.2017.02.022).

Green, B. G. (1986). Menthol inhibits the perception of warmth. *Physiol. Behav.*, **38**(6), 833–838.

Green, B. G. (1991). Interactions between chemical and thermal cutaneous stimuli inhibition (counter-irritation) and integration. *Somatosens. Mot. Res.*, **8**(4), 301–312.

Green, B. G. (1992). The sensory effects of l-menthol on human skin. *Somatosens. Mot. Res.*, **9**(3), 235–244.

Green, B. G. (2005). Lingual heat and cold sensitivity following exposure to capsaicin or menthol. *Chem. Senses*, **30**, (Suppl. 1), i201–i202 (doi: 10.1093/chemse/bjh184).

Green, B. G. & McAuliffe, B. L. (2000). Menthol desensitization of capsaicin irritation: evidence of a short-term anti-nociceptive effect. *Physiol. Behav.*, **68**(5), 631–639 (doi: 10.1016/s0031-9384(99)00221-8).

Green, B. G. & Pope, J. V. (2003). Innocuous cooling can produce nociceptive sensations that are inhibited during dynamic mechanical contact. *Exp. Brain Res.*, **148**(3), 290–299 (doi: 10.1007/s00221-002-1280-9).

Green, B. G. & Schoen, K. L. (2007). Thermal and nociceptive sensations from menthol and their suppression by dynamic contact. *Behav. Brain Res.*, **176**(2), 284–291 (doi: 10.1016/j.bbr.2006.10.013).

Greffrath, W., Kirschstein, T., Nawrath, T. & Treede, R. D. (2002). Acetylsalicylic acid reduces heat responses in rat nociceptive primary sensory neurons–evidence for a new mechanism of action, *Neurosci. Letters*, **320**(1-2), 61–64 (doi: 10.1016/S0304-3940(02)00033-2).

Grimm, C., Barthmes, M. & Wahl-Schott, C. (2014). TRPML3. In: B. Nilius, V. Flockerzi (eds.) *Mammalian Transient Receptor Potential (TRP) Cation Channels.* Handb. Exp. Pharmacol., Springer, **222**, 659-674 (doi: 10. 1007/978-3-642-54215-2_26).

Guinamard, R., Bouvagnet, P., Hof, T., Liu, H., Simard, C. & Sallé, L. (2015). TRPM4 in cardiac electrical activity. *Cardiovasc. Res.*, **108**(1), 21-30 (doi: 10.1093/cvr/cvv213).

Guo, J., She, J., Zeng, W., Chen, Q., Bai, X. C. & Jiang, Y. (2017). Structures of the calcium-activated, non-selective cation channel TRPM4. *Nature*, *552*(7684), 205-209 (doi: 10.1038/nature24997).

Hantute-Ghesquier, A., Haustrate, A., Prevarskaya, N. & Lehen'kyi, V. (2018). TRPM family channels in cancer. *Pharmaceuticals* (Basel), *11*(2), E58 (doi: 10.3390/ph11020058).

Hatem, S., Attal, N., Willer, J. C. & Bouhassira, D. (2006). Psychophysical study of the effects of topical application of menthol in healthy volunteers. *Pain*, *122*(1-2), 190–196 (doi: 10.1016/j.pain/2006.01.026).

Haustrate, A., Hantute-Ghesquier, A., Prevarskaya, N. & Lehen'kyi, V. (2019) TRPV6 calcium channel regulation, downstream pathways, and therapeutic targeting in cancer. *Cell Calcium.*, *80*, 117-124 (doi: 10.1016/j.ceca.2019.04.006).

Heinricher, M. M. & Fields, H. L. (2013). Central nervous system mechanisms of pain modulation. In: S.B. McMahon, M. Kotzenburg, I. Tracey, D.C. Turk (eds.) *Wall and Melzack's Textbook of Pain*. 6th ed., Elsevier, pp. 129-142.

Hellmich, U. A. & Gaudet, R. (2014). Structural biology of TRP channels. In: B. Nilius, V. Flockerzi (eds.) *Mammalian Transient Receptor Potential (TRP) Cation Channels*, Handb. Exp. Pharmacol., Springer, *223*, 963-990 (doi: 10.1007/978-3-319-05161-1_10).

Hensel, H. & Zotterman, Y. (1951). The effect of menthol on the thermoreceptors. *Acta Physiol. Scand.*, *24*(1), 27–34 (doi: 10.1111/j.1748-1716.1951.tb00824.x).

Himmel, N. J. & Cox, D. N. (2017). Sensing the cold: TRP channels in thermal nociception. *Channels* (Austin), *11*(5), 370-372 (doi: 10.1080/19336950.2017.1336401).

Himmel, N. J., Patel, A. A. & Cox, D. N. (2017). The TRP channels Pkd2, NompC, and Trpm act in cold-sensing neurons to mediate unique aversive behaviors to noxious cold in drosophila. Invertebrate nociception. *Oxford Research Encyclopedia of Neuroscience*; Oxford Univ. Press (doi: 10.1093/acrefore/9780190264086.013.166).

Hinman, A., Chuang, H. H., Bautista, D. M. & Julius, D. (2006). TRP channel activation by reversible covalent modification. *Proc. Natl.*

Acad. Sci. USA, **103**(51), 19564–19568 (doi: 10.1073/pnas. 0609598103).

Hoffmann, T., Kistner, K., Miermeister, F., Winkelmann, R., Wittmann, J., Fischer, M. J., Weidner, C. & Reeh, P. W. (2013). TRPA1 and TRPV1 are differentially involved in heat nociception of mice. *Eur. J. Pain*, **17**(10), 1472–1482 (doi: 10.1002/j.1532-2149.2013.00331.x).

Hu, H., Tian, J., Zhu, Y., Wang, C., Xiao, R., Herz, J. M., Wood, J. D. & Zhu, M. X. (2010). Activation of TRPA1 channels by fenamate non-steroidal anti-inflammatory drugs. *Pflugers Arch.*, **459**(4), 579-592 (doi: 10.1007/s00424-009- 0749-9).

Hulse, R. E., Li, Z., Huang, R. K., Zhang, J. & Clapham, D. E. (2018). Cryo-EM structure of the polycystin 2-l1 ion channel. *eLife*, **7**, e36931 (doi: 10.7554/eLife.36931).

Huynh, K. W., Cohen, M. R., Chakrapani, S., Holdaway, H. A., Stewart, P. L. & Moiseenkova-Bell, V. Y. (2014). Structural insight into the assembly of TRPV channels. *Structure*, **22**(2), 260-268 (doi: 10.1016/j.str.2013).

Inoue, N., Ito, S., Nogawa, M., Tajima, K. & Kyoi, T. (2012). Etodolac blocks the allyl isothiocyanate-induced response in mouse sensory neurons by selective TRPA1 activation, *Pharmacology*, **90**(1-2), 47–54 (doi: 10.1159/ 000338756).

Inoue, N., Ito, S., Tajima, K., Nogawa, M., Takahashi, Y., Sasagawa, T., Nakamura, A. & Kyoi, T. (2009). Etodolac attenuates mechanical allodynia in a mouse model of neuropathic pain. *J. Pharmacol. Sci.*, **109**(4), 600–605.

Ito, S., Tajima, K., Nogawa, M., Inoue, N., Kyoi, T., Takahashi, Y., Sasagawa, T., Nakamura, A., Kotera, T., Ueda, M., Yamashita, Y. & Banno, K. (2012). Etodolac a cyclooxygenase-2 inhibitor attenuates paclitaxel-induced peripheral neuropathy in a mouse model of mechanical allodynia. *J. Pharmacol. Exp. Ther.*, **342**(1), 53–60 (doi: 10.1124/jpet.111.187401).

Irie, S. & Furukawa, T. (2014). TRPM1. In: B. Nilius, V. Flockerzi (eds.) *Mammalian Transient Receptor Potential (TRP) Cation Channels.*

Handb. Exp. Pharmacol., Springer, *222*, 387-402 (doi: 10.1007/978-3-642-54215-2_15).

Jeon, S. & Caterina, M. J. (2018). Molecular basis of peripheral innocuous warmth sensitivity. Chapter 4, *Handb. Clin. Neurol.*, Elsevier, *156*, 69-82 (doi: 10.1016/B978-0-444-63912-7.00004-7).

Jin, P., Bulkley, D., Guo, Y., Zhang, W., Guo, Z., Huynh, W., Wu, S., Meltzer, S., Cheng, T., Jan, L. Y., Jan Y. N. & Cheng, Y. (2017). Electron cryo-microscopy structure of the mechanotransduction channel NOMPC. *Nature, 547* (7661), 118-122 (doi: 10.1038/nature 22981).

Jordt, S. E., Bautista, D. M., Chuang, H. H., McKemy, D. D., Zygmunt, P. M., Högestätt, E. D., Meng, I. D. & Julius, D. (2004). Mustard oils and cannabinoids excite sensory nerve fibres trough the TRP channel ANKTM1. *Nature, 427*(6971), 260-265 (doi: 10.1038/nature02282).

Jordt, S. E., McKemy, D. D. & Julius, D. (2003). Lessons from peppers and peppermint: the molecular logic of thermo-sensation. *Curr. Opin. Neurobiol.*, *13*(4), 487–492.

Journigan, V. B. & Zaveri, N. T. (2013). TRPM8 ion channel ligands for new therapeutic applications and as probes to study menthol pharmacology. *Life Sci.*, *92*(8-9), 425-437 (doi: 10.1016/j.lfs.2012.10. 032).

Julius, D. (2013). TRP channels and pain. *Annu. Rev. Cell. Dev. Biol.*, *29*, 355-384 (doi: 10.1146/annurev-cellbio-101011-155833).

Kaneko, Y. & Szallasi, A. (2014). Transient receptor potential (TRP) channels: a clinical perspective. *Brit. J. Pharmacol.*, *171*(10), 2474-2507 (doi: 10.1111/bph).

Kang, D., Choe, C., Cavanaugh, E., & Kim, D. (2007) Properties of single two-pore domain TREK-2 channels expressed in mammalian cells. J. Physiol., *583*(Pt. 1), 57–69 (doi: 10.1113/ jphysiol.2007.136150)

Kang, D., Choe, C. & Kim, D. (2005). Thermosensitivity of the two-pore domain K^+ channels TREK-2 and TRAAK. *J. Physiol.*, *564*(Pt. 1), 103–116 (doi: 10.1113/jphysiol.2004.081059).

Karashima, Y., Damann, N., Prenen, J., Talavera, K., Segal, A., Voets, T. & Nilius, B. (2007). Bimodal action of menthol on the transient receptor potential channel TRPA1. *J. Neurosci.*, **27**(37), 9874–9884 (doi: 10.1523/JNEUROSCI.2221-07.2007).

Karashima, Y., Talavera, K., Everaerts, W., Janssens, A., Kwan, K. Y., Vennekens, R., Nilius, B. & Voets, T. (2009). TRPA1 acts as a cold sensor *in vitro* and *in vivo*. *Proc. Natl. Acad. Sci. USA*, **106**(4), 1273-1278 (doi: 10.1073/pnas.0808487106).

Kashio, M., Sokabe, T., Shintaku, K., Uematsu, T., Fukuta, N., Kobayashi, N., Mori, Y. & Tominaga, M. (2012). Redox signal-mediated sensitization of transient receptor potential melastatin 2 (TRPM2) to temperature affects macrophage functions. *Proc. Natl. Acad. Sci. USA*, **109**(17), 6745-6750 (10.1073/pnas.1114193109).

Kashio, M. & Tominaga, M. (2017). The TRPM2 channel: A thermosensitive metabolic sensor. *Channels*, **11**(5), 426–433 (doi: 10.1080/19336950.2017.1344801).

Kerstein, P., del Camino, D., Moran, M. & Stucky, C. (2009). Pharmacological blockade of TRPA1 inhibits mechanical firing in nociceptors. *Mol. Pain*, **5,** art. 19 (doi: 10.1186/1744-8069-5-19).

Klein, A. H., Sawyer, C. M., Iodi Carstens, M., Tsagareli, M. G., Tsiklauri, N. & Carstens, E. (2010). Topical application of L-menthol induces heat analgesia, mechanical allodynia, and a biphasic effect on cold sensitivity in rats. *Behav. Brain Res.*, **212**(2), 179-186 (doi: 10.1016/j.bbr.2010.04.015).

Kobayashi, K., Fukuoka, T., Obata, K., Yamanaka, H., Dai, Y., Tokunaga, A. & Noguchi, K. (2005). Distinct expression of TRPM8, TRPA1 and TRPV1 mRNAs in rat primary afferent neurons with A-delta/C-fibers and colocalization with trk receptors. *J. Comp. Neurol.*, **493**(4), 596-606 (doi: 10.1002/cne.20794).

Kojima, I. & Nagasava, M. (2014). TRPV2. In: B. Nilius, V. Flockerzi (eds.) *Mammalian Transient Receptor Potential (TRP) Cation Channels.* Handb. Exp. Pharmacol., Springer, **222**, 247-272 (doi: 10.1007/978-3-642-54215-2_10).

Koltzenburg, M., Lundberg, L. E. & Torebjörk, H. E. (1992). Dynamic and static components of mechanical hyperalgesia in human hairy skin. *Pain*, *51*(2), 207-219.

Kozak, J. A. & Rychkov, G. (2018). Electrophysiological methods for recording CRAC and TRPV5/6 channels. In: J. A. Kozak, Putney J. W. Jr. (eds.) *Calcium Entry Channels in Non-Excitable Cells.* Chapter 1. Boca Raton (FL): CRC Press.

Kress, M., Vyklickì, L. & Reeh, P. W. (1996). Inhibition of a capsaicin induced ionic current – a new mechanism of action for aspirin-like drugs. *Pflügers Arch.*, *431*(Suppl), R61.

Krogsaeter, E. K., Biel, M., Wahl-Schott, C. & Grimm, C. (2019). The protein interaction networks of mucolipins and two-pore channels. *Biochim. Biophys. Acta Mol. Cell. Res.*, *1866*(7), 1111-1123 (doi: 10.1016/j.bbamcr.2018. 10.020).

Kwan, K. Y., Allchorne, A. J., Vollrath, M. A., Christensen, A., Zhang, D. S., Woolf, C. J. & Corey, D. P. (2006). TRPA1 contributes to cold, mechanical and chemical nociception but is not essential for hair-cell transduction. *Neuron*, *50*(2), 277-289.

Lamas, J. A., Rueda-Ruzafa, L. & Herrera-Pérez, S. (2019). Ion channels and thermosensitivity: TRP, TREK, or Both? *Int. J. Mol. Sci.*, *20*(10), pii: E2371 (doi: 10.3390/ijms20102371).

Latorre, R., Zaelzer, C. & Brauchi, S. (2009). Structure-functional intimacies of transient receptor potential channels. *Quart. Rev. Biophysics*, *42*(3), 201–246 (doi: 10.1017/S0033583509990072).

Laursen, W. J., Anderson, E. O., Hoffstaetter, L. J., Bagriantsev, S. N. & Gracheva, E. O. (2015). Species-specific temperature sensitivity of TRPA1. *Temperature* (Austin), *2*(2), 214-226 (doi: 10.1080/23328940.2014.1000702).

Li, H. (2017). TRP classification. In: Y. Wang (ed.) *Transient Receptor Potential Canonical Channels and Brain Diseases. Adv. Exp. Med. Biol.*, Springer, *976*, 1-8 (doi: 10.1007/978-94-024-1088-4_1).

Li, H., Sun, S., Chen, J., Xu, G., Wang, H. & Qian, Q. (2017). Genetics of magnesium disorders. *Kidney Dis.*, (Basel)., *3*(3), 85-97 (doi: 10. 1159/000477730).

Li, J., Zhang, X., Song, X., Liu, R., Zhang, J. & Li, Z. (2019). The structure of TRPC ion channels. *Cell Calcium*, **80**, 25-28 (doi: 10.1016/j.ceca.2019.03.005).

Li, M., Yu, Y. & Yang, J. (2011). Structural biology of TRP channels. In: M.S. Islam (ed.) Transient Receptor Potential Channels. Chapter 1. *Adv. Exp. Med. Biol.*, Springer, **704**, 1-23 (doi: 10.1007/978-94-007-0265-3_1).

Lichtenegger, M. & Groschner, K. (2014). TRPC3: A multifunctional signaling molecule. In: B. Nilius, V. Flockerzi (eds.) Mammalian Transient Receptor Potential (TRP) Cation Channels. *Handb. Exp. Pharmacol.*, Springer, **222**, 67-84 (doi: 10.1007/978-3-642-54215-2_4).

Liedtke, W. B. (2017). TRPV channels' function in osmo- and mechano-transduction. In: W.B. Liedtke, S. Heller (eds.) *TRP Ion Channel Function in Sensory Transduction and Cellular Signaling Cascades*. Chapter 22, Boca Raton (FL.), CRC Press.

Liman, E. R. (2014). TRPM5. In: B. Nilius, V. Flockerzi (eds.) *Mammalian Transient Receptor Potential (TRP) Cation Channels,* Handb. Exp. Pharmacol., Springer, **222**, 489-502 (doi: 10.1007/978-3-642-54215-2_19).

Liu, Y., Lu, F., Jiang, H., & Tang, Y. (2017) Positive selection acted on the extracellular transmembrane linkers of heat receptors during evolution. *J. Therm. Biol.*, 64: 86-91 (doi: 10.1016/j.jtherbio.2016.12.004)

Lolignier, S., Gkika, D., Andersson, D., Leipold, E., Vetter, I., Viana, F., Noël, J. & Busserolles, J. (2016). New insight in cold pain: role of ion channels, modulation, and clinical perspectives. *J Neurosci.*, **36**(45), 11435-11439 (doi: 10.1523/JNEUROSCI.2327-16.2016).

LaMotte, R. H., Lundberg, L. E. & Torebjörk, H. E. (1992). Pain, hyperalgesia and activity in nociceptive C units in humans after intradermal injection of capsaicin. *J. Physiol.*, **448**, 749–764.

LaMotte, R. H., Shain, C. N., Simone, D. A. & Tsai, E. F. (1991). Neurogenic hyperalgesia: psychophysical studies of underlying mechanisms. *J. Neurophysiol.*, **66**(1), 190–211.

Macpherson, L. J., Dubin, A. E., Evans, M. J., Marr, F., Schultz, P. G., Cravatt, B. F. & Patapoutian, A. (2007). Noxious compounds activate TRPA1 ion channels through covalent modification of cysteines. *Nature*, *445*(7127), 541–545 (doi: 10.1038/nature05544).

Macpherson, L. J., Geierstanger, B. H., Viswanath, V., Bandell, M., Eid, S. R., Hwang, S. & Patapoutian, A. (2005). The pungency of garlic: activation of TRPA1 and TRPV1 in response to allicin. *Curr. Biol.*, *15*(10), 929–934 (doi: 10.1016/j.cub.2005.04.018).

Macpherson, L. J., Hwang, S. W., Miyamoto, T., Dubin, A. E., Patapoutian, A. & Story, G. M. (2006). More than cool: promiscuous relationships of menthol and other sensory compounds. *Mol. Cell. Neurosci.*, *32*(4), 335–343 (doi: 10.1016/j.mcn.2006.05.005).

Madej, M. & Ziegler, C. M. (2018). Dawning of a new era in TRP channel structural biology by cryo-electron microscopy. *Flugers Arch.*, *470*(2), 213-225 (doi: 10.1007/s00424-018-2107-2).

Madrid R., Donovan-Rodríguez T., Meseguer V., Acosta M.C., Belmonte C., Viana F. (2006) Contribution of TRPM8 channels to cold transduction in primary sensory neurons and peripheral nerve terminals. *J. Neuroscience,* 26 (48): 12512-12525 (doi: 10.1523/JNEUROSCI.1752-06.2006)

Maeda, T., Suzuki, A., Koga, K., Miyamoto, C., Maehata, Y., Ozawa, S., Hata, R. I., Nagashima, Y., Nabeshima, K., Miyazaki, K. & Kato, Y. (2017). TRPM5 mediates acidic extracellular pH signaling and TRPM5 inhibition reduces spontaneous metastasis in mouse B16-BL6 melanoma cells. *Oncotarget*, *8*(45), 78312–78326 (doi: 10.18632/oncotarget.20826).

Magnusson, B. M., Walters, K. A. & Roberts, M. S. (2001). Veterinary drug delivery: potential for skin penetration enhancement. *Adv. Drug Deliv. Rev.*, *50*(3), 205–227.

Maher, M., Ao, H., Banke, T., Nasser, N., Wu, N. T., Breitenbucher, J. G., Chaplan, S. R. & Wickenden, A. D. (2008). Activation of TRPA1 by farnesyl thiosalicylic acid. *Mol. Pharmacol.*, *73*(4), 1225-1234 (doi: 10.1124/mol. 107.042663).

Malkia, A., Madrid, R., Meseguer, V., de la Peña, E., Valero, M., Belmonte, C. & Viana, F. (2007). Bidirectional shifts of TRPM8 channel gating by temperature and chemical agents modulate the cold sensitivity of mammalian thermoreceptors. *J. Physiol.*, *581*(Pt 1), 155–174 (doi: 10.1113/jphysiol.2006.123059).

Marshall, L. J., Austin, C. P., Casey, W., Fitzpatrick, S. C., Willett, C. (2018). Recommendations toward a human pathway-based approach to disease research. *Drug Discov. Today*, *23*(11), 1824-1832 (doi: 10.1016/j. drudis. 2018.05.038).

Marthar, I., Jacobs, G., Kecskes, M., Menigoz, A., Philippaert, K. & Vennekens, R. (2014). TRPM4. In: B. Nilius, V. Flockerzi (eds.) *Mammalian Transient Receptor Potential (TRP) Cation Channels.* Handb. Exp. Pharmacol., Springer, *222*, 461-487 (doi: 10.1007/978-3-642-54215-2_18).

Martínez, A. L., González-Trujano, M. E., Chávez, M., Pellicer, F., Moreno, J. & López-Muñoz, F. J. (2011). Hesperidin produces antinociceptive response and synergistic interaction with ketorolac in an arthritic gout-type pain in rats. *Pharmacol. Biochem. Behav.*, *97*(4), 683-689 (doi: 10.1016/j.pbb.2010.11.010).

Materazzi, S., Nassini, R., Andrè, E., Campi, B., Amadesi, S., Trevisani, M., Bunnett, N. W., Patacchini, R. & Geppetti, P. (2008). Cox-dependent fatty acid metabolites cause pain through activation of the irritant receptor TRPA1. *Proc. Natl. Acad. Sci. USA*, *105*(33), 12045-12050 (doi: 10.1073/pnas.0802354105).

Matsuura, H., Sokabe, T., Kohno, K., Tominaga, M. & Kadowaki, T. (2009). Evolutionary conservation and changes in insect TRP channels. *BMC Evol. Biol.*, *9*, 228 (doi: 10.1186/1471-2148-9-228).

Maurer, K., Binzena, U., Mörza, H., Bugert, P., Schedel, A., Treede, R. D. & Greffrath, W. (2014). Acetylsalicylic acid enhances tachyphylaxis of repetitive capsaicin responses in TRPV1-GFP expressing HEK293 cells. *Neurosci. Letters*, *563*, 101–106 (doi: 10.1016/j.neulet.2014.01.050).

Mederos y Schnitzler, M., Gudermann, T. & Storch, U. (2018). Emerging roles of diacylglycerol-sensitive TRPC4/5 channels. *Cells*, *7*(11). pii: E218 (doi: 10.3390/cells7110218).

Merrill, A. W., Cuellar, J. M., Judd, J. H., Iodi Carstens, M. & Carstens, E. (2008). Effects of TRPA1 agonists mustard oil and cinnamaldehyde on lumbar spinal wide-dynamic range neuronal responses to innocuous and noxious cutaneus stimuly in rats. *J. Neurophysiol.*, *99*(2), 415-425 (doi: 10.1152/jn.00883.2007).

Meotti, F. C., de Andrade, E. L. & Calixto, J. B. (2014). TRP modulation by natural compounds. In: B. Nilius, V. Flockerzi (eds.) *Mammalian Transient Receptor Potential (TRP) Cation Channels*, Handb. Exp. Pharmacol., Springer, *223*, 1177-1238 (doi: 10.1007/978-3-319-05161-1_19).

Miller, B. A. (2014). TRPC2. In: B. Nilius, V. Flockerzi (eds.) *Mammalian Transient Receptor Potential (TRP) Cation Channels*, Handb. Exp. Pharmacol., Springer, *222*, 53-65 (doi: 10.1007/978-3-642-54215-2_3).

Miller, B. A. (2019). TRPM2 in cancer. *Cell Calcium*, *80*, 8-17 (doi: 10.1016/j.ceca.2019.03.002).

Mishra, S. K., Tisel, S. M., Orestes, P., Bhangoo, S. K. & Hoon, M. A. (2011). TRPV1-lineage neurons are required for thermal sensation. *EMBO J.*, *30*(3), 582–593 (doi: 10.1038/emboj.2010.325).

Montell, C. (2001). Physiology, phylogeny, and functions of the TRP superfamily of cation channels. *Science's STKE*, *90*, re1 (doi: 10.1126/stke.2001.90.re1).

Montell, C. & Rubin, G. M. (1989). Molecular characterization of the Drosophila trp locus: a putative integral membrane protein required for phototransduction. *Neuron*, *2*(4), 1313-1323.

Moore, C. & Liedtke, W. B. (2017). Osmomechanical-sensitive TRPV channels in mammals. In: T.L.R. Emir (ed.) *Neurobiology of TRP Channels*. 2nd edition, chapter 5. Boca Raton (FL): CRC Press/Taylor & Francis (doi: 10.4324/9781315152837-5).

Moqrich, A., Hwang, S. W., Earley, T. J., Petrus, M. J., Murray, A. N., Spencer, K. S. R., Andahazy, M., Story, G. M. & Patapoutian A.

(2005). Impaired thermosensation in mice lacking TRPV3, a heat and camphor sensor in the skin. *Science*, *307*(5714), 1468–1472 (doi: 10.1126/science.1108609).

Moran, M. M. (2018). TRP channels as potential drug targets. *Annu. Rev. Pharmacol. Toxicol.*, *58*, 309–330 (doi: 10.1146/annurev-pharmtox-010617-052832).

Moran, M. M., McAlexander, M. A., Bíró, T. & Szallasi, A. (2011). Transient receptor potential channels as therapeutic targets. *Nature Rev. Drug Discovery*, *10*(8), 601-620 (doi: 10.1038/nrd3456).

Moran, M. M. & Szallasi, A. (2018). Targeting nociceptive transient receptor potential channels to treat chronic pain: Current state of the field. *Brit. J. Pharmacol.*, *175*(12), 2185–2203 (doi: 10.1111/bph.14044).

Morrison, S. F. & Nakamura, K. (2011). Central neural pathways for thermoregulation. *Front. Biosci.* (Landmark Ed.), *16*, 74-104.

McKemy, D. D., Neuhausser, W. M. & Julius, D. (2002). Identification of a cold receptor reveals a general role for TRP channels in thermosensation. *Nature*, *416*(6876), 52–58 (doi: 10.1038/nature719).

McNamara, C. R., Mandel-Brehm, J., Bautista, D. M., Siemens, J., Deranian, K. L., Zhao, M., Hayward, N. J., Chong, J. A., Julius, D., Moran, M. M. & Fanger, C. M. (2007). TRPA1 mediates formalin-induced pain. *Proc. Natl. Acad. Sci. USA*, *104*(33), 13525-13530 (doi: 10.1073/pnas.0705924104).

Munns, C., Al Qatari, M. & Koltzenburg, M. (2007). Many cold sensitive peripheral neurons of the mouse do not express TRPM8 or TRPA1. *Cell Calcium*, *41*(4), 331–342 (doi: 10.1016/j.ceca.2006.07.008).

Na, T. & Peng, J. B. (2014). TRPV5: A Ca^{2+} channel for the fine-tuning of Ca^{2+} reabsorption. In: B. Nilius, V. Flockerzi (eds.) *Mammalian Transient Receptor Potential (TRP) Cation Channels*. Handb. Exp. Pharmacol., Springer, *222*, 321-357 (doi: 10.1007/978-3-642-54215-2_13).

Nakamura, K. & Morrison, S. F. (2011). Central efferent pathways for cold-defensive and febrile shivering. *J. Physiol.*, *589*(Pt 14), 3641-3658 (doi: 10.1113/jphysiol.2011.210047).

Namer, B., Seifert, F., Handwerker, H. O. & Maihöfner, C. (2005). TRPA1 and TRPV1 activation in humans: effects of cinnamaldehyde and menthol. *Neuroreport*, *16*(9), 955-959 (doi: 10.1097/00001756-20050 6210-00015).

Nassini, R., Materazzi, S., Benemei, S. & Geppetti, P. (2014). The TRPA1 channel in inflammatory and neuropathic pain and migraine. *Rev. Physiol. Biochem. Pharmacol.*, *176*, 1–43 (doi: 10.1007/112_2014 _18).

Nassini, R., Fusi, C., Materazzi, S., Coppi, E., Tuccinardi, T., Marone, I. M., De Logu, F., Preti, D., Tonello, R., Chiarugi, A., Patacchini, R., Geppetti, P. & Benemei, S. (2015). The TRPA1 channel mediates the analgesic action of dipyrone and pyrazolone derivatives. *Brit. J. Pharmacol.*, *172*(13), 3397-3411 (doi: 10.1111/bph. 13129).

Navratilova, E., Xie, J. Y., King, T. & Porreca, F. (2013). Evaluation of reward from pain relief. *Ann. NY Acad. Sci.*, *1282*, 1–11 (doi: 10.1111/nyas.12095).

Navratilova, E., Xie, J. Y., Okun, A., Qu, C., Eyde, N., Ci, S., Ossipov, M.H., King, T., Fields, H.L. & Porreca, F. (2012). Pain relief produces negative reinforcement through activation of mesolimbic reward-valuation circuitry. *Proc. Natl. Acad. Sci. USA*, *109*(50), 20709–20713 (doi: 10.1073/pnas.1214605109).

Nealen, M. L., Gold, M. S., Thut, P. D. & Caterina, M. J. (2003). TRPM8 mRNA is expressed in a subset of cold-responsive trigeminal neurons from rat. *J. Neurophysiol.*, *90*(1), 515–520 (doi: 10.1152/jn.00843.2002).

Neddermeyer, T. J., Flühr, K. & Lötsch, J. (2008). Principle components analysis of pain thresholds to thermal, electrical, and mechanical stimuli suggests a predominant common source of variance. *Pain*, *138*(2), 286–291 (doi: 10.1016/j.pain.2007.12.015).

Nedungadi, T. P., Dutta, M., Bathina, C. S., Caterina, M. J. & Cunningham, J. T. (2012). Expression and distribution of TRPV2 in rat brain. *Exp. Neurol.*, *237*(1), 223–237 (doi: 10.1016/j.expneurol.2012.06.017).

Nesin, V. & Tsiokas, L. (2014). TRPC1. In: B. Nilius, V. Flockerzi (eds.) *Mammalian Transient Receptor Potential (TRP) Cation Channels.* Handb. Exp. Pharmacol., Springer, **222**, 15-51 (doi: 10.1007/978-3-642-54215-2_2).

Nilius, B. & Appendino, G. (2013). Spices: the savory and beneficial science of pungency. *Rev. Physiol. Biochem. Pharmacol.*, (Springer), **164**, 1-76 (doi: 10.1007/112_2013_11).

Nilius, B., Appendino, G. & Owsianik, G. (2012). The transient receptor potential channel TRPA1: from gene to pathophysiology. Pflugers Arch. **464**(5):425-458 (doi: 10.1007/s00424-012-1158-z).

Nilius, B. & Bíró, T. (2013). TRPV3: a 'more than skinny' channel. *Exp. Dermatol.*, **22**(7): 447-452 (doi: 10.1111/exd.121630).

Nilius, B., Bíró, T. & Owsianik, G. (2014). TRPV3: time to decipher a poorly understood family member! *J. Physiol.*, **592**(2), 295-304 (doi: 10.1113/jphysiol.2013.255968).

Nilius, B. & Flockerzi, V. (2014). What do we really know and what do we need to know: Some controversies, perspectives, and surprises. In: B. Nilius, V. Flockerzi (eds.) *Mammalian Transient Receptor Potential (TRP) Cation Channels.* Handb. Exp. Pharmacol., Springer, **223**, 1239-1280 (doi: 10.1007/978-3-319-05161-1_20).

Nilius, B. & Owsianik, G. (2010). Transient receptor potential channelopathies. *Pflügers Archive*, **460**(2), 437-50 (doi:10.1007/s00 424-010-0788-2).

Nilius, B. & Szallasi, A. (2014). Transient receptor potential channels as drug targets: from the science of basic research to the art of medicine. *Pharmacol. Rev.*, **66**(3), 676–814 (doi:_10.1124/pr.113.008268).

Noben-Trauth, K. (2011). The TRPML3 channel: from gene to function. In: M.S. Islam (ed.) *Transient Receptor Potential Channels.* Chapter 13. Adv. Exp. Med. Biol., Springer, **704**, 229-237 (doi: 10.1007/978-94-007-0265-3_13).

Noel, J., Zimmermann, K., Busserolles, J., Deval, E., Alloui, A., Diochot, S., Guy, N., Borsotto, M., Reeh, P., Eschalier, A. & Lazdunski, M. (2009). The mechano-activated K^+ channels TRAAK and TREK-1 control both warm and cold perception. *EMBO J.*, *28*(9), 1308–1318 (doi: 10.1038/emboj.2009.57).

Noyer, L., Grolez, G. P., Prevarskaya, N., Gkika, D. & Lemonnier, L. (2018). TRPM8 and prostate: a cold case? *Pflugers Arch.*, *470*(10), 1419-1429 (doi: 10.1007/s00424-018-2169-1).

Nozadze. I., Tsiklauri. N., Gurtskaia. G., Abzianidze. E. & Tsagareli, M. G. (2016a). TRP channels in thermal pain sensation. In: *Systemic, Cellular and Molecular Mechanisms of Physiological Functions and Their Disorders.* Chapter 22. New York: Nova Biomedical, pp. 271-287.

Nozadze, I., Tsiklauri, N., Gurtskaia, G. & Tsagareli, M. G. (2016b). Role of thermo TRPA1 and TRPV1 channels in heat, cold, and mechanical nociception of rats. *Behav. Pharmacol.*, *27*(1), 29-36 (doi:10.1097/FBP.0000000000000176).

Nozadze, I., Tsiklauri, N., Gurtskaia, G. & Tsagareli, M. G. (2016c). NSAIDs attenuate hyperalgesia induced by TRP channel activation. *Data Brief*, *6*, 668-673 (doi:10.1016/j.dib.2015.12.055).

Nozadze, I., Tsiklauri, N., Gurtskaia, G. & Tsagareli, M. G. (2018). NSAIDs attenuate agonist-evoked activation of the TRPA1 channel: Behavioral evidence. In: *Systemic, Cellular and Molecular Mechanisms of Physiological Functions and Their Disorders*. New York: Nova, pp. 257-271.

Nozadze, I., Tsiklauri, N., Gurtskaia, G. & Tsagareli, M. G. (2019). Agonist-evoked hyperalgesia and allodynia: The role of transient receptor potential channels in pain and itch. In: M.G. Tsagareli (ed.) *Hyperalgesia and Allodynia: A Closer Look. Symptoms, Mechanisms and Treatment.* Chapter 7, New York: Nova, pp. 151-174.

Oberwinkler, J. & Philipp, S. E. (2014). TRPM3. In: B. Nilius, V. Flockerzi (eds.) *Mammalian Transient Receptor Potential (TRP) Cation Channels.* Handb. Exp. Pharmacol., *222*, 427-459 (doi: 10.1007/978-3-642-54215-2_17).

Ogunrinde, A., Pereira, R. D., Beaton, N., Lam, D. H. Whetstone, C. & Hill, C. E. (2017). Hepatocellular differentiation status is characterized by distinct subnuclear localization and form of the chanzyme TRPM7. *Differentiation*, **96**, 15–25 (doi: 10.1016/j.diff.2017.06.00).

Ong, H. L., de Souza, L. B., Cheng, K. T. & Ambudkar, I. S. (2014). Physiological functions and regulation of TRPC channels. In: B. Nilius, V. Flockerzi (eds.) *Mammalian Transient Receptor Potential (TRP) Cation Channels.* Handb. Exp. Pharmacol., Springer, **223**, 1005-1034 (doi: 10.1007/978-3-319-05161-1_12).

Ordás, P., Hernández-Ortego, P., Vara, H., Fernández-Peña, C., Reimúndez, A., Morenilla-Palao, C., Guadaño-Ferraz, A., Gomis, A., Hoon, M., Viana, F. & Señarís, R. (2019). Expression of the cold thermoreceptor TRPM8 in rodent brain thermoregulatory circuits. *J. Comp. Neurol.* (doi: 10.1002/cne.24694).

Parenti, A., De Logu, F., Geppetti, P. & Benemei, S. (2016). What is the evidence for the role of TRP channels in inflammatory and immune cells? *Brit. J. Pharmacol.*, **173**(6), 953-969 (doi: 10.1111/bph.13392).

Paulsen, C. E., Armache, J. P., Gao, Y., Cheng, Y. & Julius, D. (2015). *Structure of the TRPA1 ion channel suggests regulatory mechanisms. Nature,* **520**(7548), 511-517 (doi: 10.1038/nature14367).

Peier, A. M., Moqrich, A., Hergarden, A. C., Reeve, A. J., Andersson, D. A., Story, G. M., Early, T. J., Dragoni, I., McIntyre, P., Bevan, S. & Patapoutian, A. (2002). A TRP channel that senses cold stimuli and menthol. *Cell*, **108**(5), 705–715 (doi: 10.1016/s0092-8674(02)00652-9).

Peng, J. B., Suzuki, Y., Gyimesi, G. & Hediger, M. A. (2018). TRPV5 and TRPV6 calcium-selective channels. In: Kozak J.A., Putney JW Jr. (eds.) *Calcium Entry Channels in Non-Excitable Cells.* Chapter 13. Boca Raton (FL): CRC Press.

Pérez de Vega, M. J., Ferrer-Montiel, A. & González-Muñiz, R. (2018). Recent progress in non-opioid analgesic peptides. *Arch. Biochem. Biophys.*, **660**(1), 36-52 (doi: 10.1016/j.abb.2018.10.011).

Pergolizzi, J. V., Taylor, R., LeQuang, J. A. & Raffa, R. B. (2018). The role and mechanism of action of menthol in topical analgesic products. *J. Clin. Pharm. Ther.*, **43**(3), 313-319 (doi: 10.1111/jcpt.12679).

Pertovaara, A. & Koivisto, A. (2011). TRPA1 ion channel in the spinal dorsal horn as a therapeutic target in central pain hypersensitivity and cutaneous neurogenic inflammation. *Eur. J. Pharmacol.*, *666*(1-3), 1–4 (doi: 10.1016/j.ejphar.2011.05.027).

Petrus, M., Peier, A. M., Bandell, M., Hwang, S. W., Huynh, T., Olney, N., Jegla, T. & Patapoutian, A. (2007). A role of TRPA1 in mechanical hyperalgesia is revealed by pharmacological inhibition. *Mol. Pain*, *3*, art. 40 (doi:10. 1186/1744-8069-3-40).

Proudfoot, C. J., Garry, E. M., Cottrell, D. F., Rosie, R., Anderson, H., Robertson, D. C., Fleetwood-Walker, S. M. & Mitchell, R. (2006). Analgesia mediated by the TRPM8 cold receptor in chronic neuropathic pain. *Curr. Biol.*, *16*(16), 1591–605 (doi: 10.1016/j.cub.2006.07.061).

Pyrski, M., Eckstein, E., Schmid, A. Bufe, B., Weiss, J., Chubanov, V., Boehm, U. & Zufall, F. (2017). Trpm5 expression in the olfactory epithelium. *Mol. Cell. Neurosci.*, *80*, 75-88 (doi: 10.1016/j.mcn.2017.02.002).

Radresa, O., Paré, M. & Albert, J. S. (2013). Multiple roles of transient receptor potential (TRP) channels in inflammatory conditions and current status of drug development. *Curr. Top. Med. Chem.*, *13*(3), 367-385.

Reid, G. (2005). ThermoTRP channels and cold sensing: what are they really up to? *Pflugers Arch.*, *451*(1), 250–263 (doi: 10.1007/s00424-005-1437-z).

Reid, G., Babes, A. & Pluteanu, F. (2002). A cold- and menthol-activated current in rat dorsal root ganglion neurons: properties and role in cold transduction. *J. Physiol.*, *545*(2), 595–614 (doi: 10.1113/jphysiol.2002.024331).

Raouf, R., Quick, K. & Wood, J. N. (2010). Pain as a channelopathy. *J. Clin. Invest.*, *120*(11), 3745-3752 (doi: 10.1172/JCI43158).

Reid, G. & Flonta, M. L. (2001). Cold transduction by inhibition of a background potassium conductance in rat primary sensory neurons. *Neurosci. Letters*, *297*(3), 171–174 (10.1016/s0304-3940(00)01694-3).

Ren, K. & Dubner, R. (2009). Descending control mechanisms. In: A.I. Basbaum, M.C. Bushnell (eds.) *Science of Pain.* Elsevier, pp. 723-762.

Rohacs, T., Lopes, C. M. B., Michailidis, I. & Logothetis, D. E. (2005). PI(4,5)P-2 regulates the activation and desensitization of TRPM8 channels through the TRP domain. *Nature Neurosci., 8*(5), 626–634 (doi: 10.1038/nn1451).

Rose, T. M., Reilly, C. A., Deering-Rice, C. E., Brewster, C. & Brewster, C. (2014). Inhibition of FAAH, TRPV1, and COX2 by NSAID–serotonin conjugates. *Bioorg. Med. Chem. Letters, 24*(24), 5695–5698 (doi: 10.1016/j.bmcl.2014.10.064).

Rossi, H. L., Vierck, C. J., Caudle, R. M. & Neubert, J. K. (2006). Characterization of cold sensitivity and thermal preference using an operant orofacial assay. *Mol. Pain, 2*, art. 37 (doi: 10.1186/1744-8069-2-37).

Rubaiy, H. N. (2019). Treasure troves of pharmacological tools to study transient receptor potential canonical 1/4/5 channels. *Brit. J. Pharmacol., 176*(7), 832-846 (doi: 10.1111/bph.14578).

Saito, S. & Shingai, R. (2006). Evolution of thermoTRP ion channel homologs in vertebrates. *Physiol. Genomics, 27*(3), 219-230 (doi: 10.1152/physiolgenomics.00322.2005).

Samanta, A., Hughes, T. E. T. & Moiseenkova-Bell, V. Y. (2018). Transient Receptor Potential (TRP) channels. In: J.R. Harris, E.J. Boekema (eds.) *Membrane Protein Complexes: Structure and Function.* Chapt. 6. *Subcell. Biochem., 87*, 141-165 (doi: 10.1007/978-981-10-7757-9_6).

Santoni, G. & Amantini, C. (2019). The transient receptor potential vanilloid type-2 (TRPV2) ion channels in neurogenesis and gliomagenesis: Cross-talk between transcription factors and signaling molecules. *Cancers* (Basel), *11*(3), E322 (doi: 10.3390/cancers 11030322).

Sawyer, C. M., Iodi Carstens, M. & Carstens, E. (2009). Mustard oil enhances spinal neuronal responses to noxious heat but not cooling. *Neurosci. Letters, 461*(3), 271–274 (doi: 10.1016/j.neulet.2009.06.036).

Schneider, F. M., Mohr, F., Behrendt, M. & Oberwinkler, J. (2015). Properties and functions of TRPM1 channels in the dendritic tips of retinal ON-bipolar cells. *Eur. J. Cell. Biol.*, *94*(7-9), 420-427 (doi: 10.1016/j.ejcb.2015. 06. 005).

Semmo, M., Köttgen, M. & Hofherr, A. (2014). The TRPP subfamily and Polycystin-1 proteins. In: B. Nilius, V. Flockerzi (eds.) *Mammalian Transient Receptor Potential (TRP) Cation Channels*. Handb. Exp. Pharmacol., Springer, *222*, 675-711 (doi: 10.1007/978-3-642-54215-2_27).

Señarís, R., Ordás, P., Reimúndez, A. & Viana, F. (2018). Mammalian cold TRP channels: impact on thermo-regulation and energy homeostasis. *Pflugers Arch.*, *470*(5), 761-777 (doi: 10.1007/s00424-018-2145-9).

Shibasaki, K. (2016a). Physiological significance of TRPV2 as a mechanosensor, thermosensor and lipid sensor. *J. Physiol. Sci.*, *66*(5), 359-365 (doi: 10.1007/s12576-016-0434-7).

Shibasaki, K. (2016b). TRPV4 ion channel as important cell sensors. *J. Anesth.*, *30*(6), 1014-1019 (doi: 10.1007/s00540-016-2225-y).

Simons, C. T., Iodi Carstens, M. & Carstens, E. (2003). Oral irritation by mustard oil: self-desensitization and cross-desensitization with capsaicin. *Chem. Senses*, *28*(6), 459–465.

Simons, C. T., Sudo, S., Sudo, M. & Carstens, E. (2004). Mustard oil has differential effects on the response of trigeminal caudalis neurons to heat and acidity. *Pain*, *110*(1-2), 64-71 (doi: 10.1016/j.pain.2004.03. 009).

Song, K., Wang, H., Kamm, G. B., Pohle, J., Reis, F. C., Heppenstall, P., Wende, H. & Siemens, J. (2016). The TRPM2 channel is a hypothalamic heat sensor that limits fever and can drive hypothermia. *Science*, *353*(6306), 1393 –1398 (doi: 10.1126/science.aaf7537).

Spahn, V., Stein, C. & Zöllner, C. (2014). Modulation of transient receptor vanilloid 1 activity by transient receptor potential ankyrin 1. *Mol. Pharmacol.*, *85*(2), 335–344 (doi: 10.1124/mol.113.088997).

Story, G. M., Peier, A. M., Reeve, A. J., Eid, S. R., Mosbacher, J., Hricik, T. R., Earley, T. J., Hergarden, A. C., Andersson, D. A., Hwang, S.

W., McIntyre, P., Jegla, T., Bevan, S. & Patapoutian, A. (2003). ANKTM1, a TRP-like channel expressed in nociceptive neurons, is activated by cold temperatures. *Cell*, *112*(6), 819-829 (doi: 10.1016/s0092-8674(03)00158-2).

Sugibayashi, K., Kobayashi, D., Nakagaki, E., Hatanaka, T., Inoue, N., Kusumi, S., et al. (1995). Differences in enhancing effect of L-menthol, ethanol and their combination between hairless rat and human skin. *Int. J. Pharm.*, *113*, 189–197.

Sugiura, T., Tominaga, M., Katsuya, H. & Mizumura, K. (2002). Bradykinin lowers the threshold temperature for heat activation of vanilloid receptor 1. *J. Neurophysiol.*, *88*(1), 544–548 (doi: 10.1152/jn.2002.88.1.544).

Sun, H. S. (2017). Role of TRPM7 in cerebral ischemia and hypoxia. *J. Physiol.*, *595*(10), 3077-3083 (doi: 10.1113/JP273709).

Tabarean, I. V., Conti, B., Behrens, M., Korn, H. & Bartfai, T. (2005). Electrophysiological properties and thermo-sensitivity of mouse preoptic and anterior hypothalamic neurons in culture. *Neuroscience*, *135*(2), 433–449 (doi: 10.1016/j/neuroscience.2005.06.053).

Tai, Y., Yang, S., Liu, Y. & Shao, W. (2017). TRPC channels in health and disease. In: Y. Wang (ed.) *Transient Receptor Potential Canonical Channels and Brain Diseases*. Chap. 4. *Adv. Exp. Med. Biol.*, Springer, *976*, 35–45 (doi: 10.1007/978-94-024-1088-4_4).

Talavera, K., Yasumatsu, K., Voets, T., Droogmans, G., Shigemura, N., Ninomiya, Y., Margolskee, R. F. & Nilius, B. (2005). Heat activation of TRPM5 underlies thermal sensitivity of sweet taste. *Nature*, *438*(7070), 1022–1025 (doi: 10.1038/nature04248).

Tan, C. H. & McNaughton, P. A. (2018). TRPM2 and warmth sensation. *Pflügers Archiv*, *470*(5), 787-798 (doi: 10. 1007/s00424-018-2139-7).

Taylor, N. A. S. (2014). Human heat adaptation. *Compreh. Physiol.*, *4*(1), 325–365 (doi: 10.1002/cphy.c130022).

Taylor-Clark, T. E., Undem, B. J., Macglashan, D. W. Jr., Ghatta, S., Carr M. J. & McAlexander, M. A. (2008). Prostaglandin-induced activation of nociceptive neurons via direct interaction with transient receptor potential A1 (TRPA1). *Mol. Pharmacol.*, *73*(2), 274-281 (doi: 10. 1124/mol.107.040832).

Thiel, G., Rubil, S., Lesch, A., Guethlein, L. A. & Rössler, O. G. (2017). Transient receptor potential TRPM3 channels: Pharmacology, signaling, and biological functions. *Pharmacol. Res.*, *124*, 92-99 (doi: 10.1016/j.phrs.2017. 07.014).

Tipton, M. J., Pandolf, K. B., Sawka, M. N., Werner, J. & Taylor, N. A. S. (2008). Physiological adaptation to hot and cold environments. In: N.A.S. Taylor, H. Groeller (eds.) *Physiological Bases of Human Performance during Work and Exercise.* Edinburgh: Churchill Livingstone Elsevier, pp. 379–400.

Todaka, H., Taniguchi, J., Satoh, J., Mizuno, A. & Suzuki, M. (2004). Warm temperature-sensitive transient receptor potential vanilloid 4 (TRPV4) plays an essential role in thermal hyperalgesia. *J. Biol. Chem.*, *279*(34), 35133–35138 (doi: 10.1074/jbc.M406260200).

Tominaga, M. (2009). Thermal sensation (cold and heat) through thermosensitive TRP channel activation. In: A.I., Basbaum, M.C. Bushnell (eds.) *Science of Pain.* San Diego: Elsevier. pp. 127–132.

Torebjörk, H. E., Lundberg, L. E. & LaMotte, R. H. (1992). Central changes in processing of mechanoreceptive input in capsaicin-induced secondary hyperalgesia in humans. *J. Physiol.*, *448*, 765–780.

Trevisani, M., Siemens, J., Materazzi, S., Bautista, D. M., Nassini, R., Campi, B., Imamachi, N., Andrè, E., Patacchini, R., Cottrell, G. S., Gatti, R., Basbaum, A. I., Bunnett, N. W., Julius, D. & Geppetti, P. (2007). 4-Hydroxynonenal, an endogenous aldehyde, causes pain and neurogenic inflammation through activation of the irritant receptor TRPA1. *Proc. Natl. Acad. Sci. USA*, *104*(33), 13519-13524 (doi: 10. 1073/pnas.0705923104).

Tsagareli, M. G. (2011). Behavioral testing of the effects of thermo-sensitive TRP channel agonists on touch, temperature and pain sensations. *Neurophysiol.* (Springer), *43*(4), 309–320.

Tsagareli, M. G. (2013). Transient receptor potential ion channels as promising therapeutic targets: An overview. In: Atta-ur-Rahman, M. Iqbal Choudhary (eds.). *Frontiers CNS Drug Discovery*, volume 2, chapter 5. New York: Bentham, pp. 118-145 (doi: 10.2174/ 9781608057672113020007).

Tsagareli, M. G. (2015). Thermo-TRP channels in pain sensation. *Brit. J. Pharm. Sci.*, *6*(6), 376-384 (doi: 10.9734/BJPR/2015/16883).

Tsagareli, M. G., Nozadze, I., Tsiklauri, N. & Gurtskaia, G. (2018). Non-steroidal anti-inflammatory drugs attenuate agonist-evoked activation of transient receptor potential channels. *Biomed. Pharmacother.*, *97*, 745-751 (doi: 10.1016/j.biopha.2017.10.131).

Tsagareli, M. G., Nozadze, I., Tsiklauri, N. & Gurtskaia, G. (2019). TRPA1 channel is involved in SLIGRL-evoked thermal and mechanical hyperalgesia in mice. *Med. Sci.* *7*(4), art. 62 (doi: 10.3390/medsci7040062).

Tsagareli, M. G. & Tsiklauri, N. (2012). *Behavioral Study of 'Non-opioid Tolerance'*. New York: Nova Biomedical.

Tsagareli, M. G., Tsiklauri, N., Zanotto, K. L., Iodi, Carstens M., Klein, A. H., Gurtskaia, G., Sawyer, C. M., Abzianidze, E. & Carstens, E. (2010). Behavioral evidence of heat hyperalgesia and mechanical allodynia induced by intra-dermal cinnamaldehyde in rats. *Neurosci. Letters*, *473*(3), 233-236 (doi: 10.1016/j.neulet.2010.02.056).

Tsavaler, L., Shapero, M. H., Morkowski, S. & Laus, R. (2001). Trp-p8, a novel prostate-specific gene, is upregulated in prostate cancer and other malignancies and shares high homology with transient receptor potential calcium channel proteins. *Cancer Res.*, *61*, 3760–3769.

Turner, H. N., Armengol, K., Patel, A. A., Himmel, N. J., Sullivan, L., Iyer, S. C., Bhattacharya, S., Iyer, E. P. R., Landry, C., Galko, M. K. & Cox, D. N. (2016). The TRP channels Pkd2, NompC, and Trpm act in cold-sensing neurons to mediate unique aversive behaviors to noxious cold in Drosophila. *Curr. Biol.*, *26*(1), 1-13 (doi: 10.1016/j.cub.2016.09.038).

Uchida, K., Dezaki, K., Yoneshiro, T., Watanabe, T., Yamazaki, J., Saito, M., Yada, T., Tominaga, M. & Iwasaki, Y. (2017). Involvement of thermosensitive TRP channels in energy metabolism. *J. Physiol. Sci.*, *67*(5), 549-560 (doi: 10.1007/s12576-017-0552-x).

Urban, M. O., Zahn, P. K. & Gebhart, G. F. (1999). Descending facilitatory influences from the rostral medial medulla mediate secondary, but not primary hyperalgesia in the rat. *Neuroscience*, *90*(2), 349–352 (doi: 10.1016/s0306- 4522(99)00002-0).

van Goor, M. K. C., Hoenderop, J. G. J. & van der Wijst, J. (2017). TRP channels in calcium homeostasis: from hormonal control to structure-function relationship of TRPV5 and TRPV6. *BBA*, *1864*(6), 883-893 (doi: 10.1016/j.bbamcr.2016.11.027).

Vandewauw, I., De Clercq, K., Mulier, M., Held, K., Pinto, S., Van Ranst, N., Segal, A., Voets, T., Vennekens, R., Zimmermann, K., Vriens J. & Nilius B. (2018). A TRP channel trio mediates acute noxious heat sensing. *Nature*, *29*(7698), 662–666 (doi: 10.1038/nature26137).

Vangeel, L. & Voets, T. (2019). Transient receptor potential channels and calcium signaling. *Cold Spring Harb. Perspect. Biol.*, pii: a035048 (doi: 10.1101/cshperspect.a035048).

Venkatachalam, K., Luo, J. & Montell, C. (2014). Evolutionarily conserved, multitasking TRP channels: Lessons from worms and flies. In: B. Nilius, V. Flockerzi (eds.) *Mammalian Transient Receptor Potential (TRP) Cation Channels.* Handb. Exp. Pharmacol., Springer, *223*, 937-962 (doi: 10.1007/978-3-319-05161-1_9).

Venkatachalam, K. & Montell, C. (2007) TRP channels. *Annu. Rev. Biochem.*, *76*, 387–417 (doi: 10.1146/annurev.biochem.75.103004.142819).

Venkatachalam, K., Wong, C. O. & Zhu, M. X. (2015). The role of TRPMLs in endolysosomal trafficking and function. *Cell Calcium*, *58*(1), 48-56 (doi: 10.1016/j.ceca.2014.10.008).

Vennekens, R., Mesuere, M. & Philippaert, K. (2018). TRPM5 in the battle against diabetes and obesity. *Acta Physiol.* (Oxf) *222*(2), e12949 (doi: 10.1111/apha.12949).

Vetter, I., Wyse, B. D., Roberts-Thomson, S. J., Monteith, G. R. & Cabot P. J. (2008). Mechanisms involved in potentiation of transient receptor potential vanilloid 1 responses by ethanol. *Eur. J. Pain*, *12*(4), 441–454 (doi: 10.1016/j.ejpaon.2007.07.001).

Viana, F. (2016). TRPA1 channels: molecular sentinels of cellular stress and tissue damage. *J Physiol.*, *594*(15), 4151-4169 (doi: 10.1113/JP270935).

Viana, F., de la Peña, E. & Belmonte, C. (2002). Specificity of cold thermotransduction is determined by differential ionic channel expression. *Nature Neurosci.*, *5*(3), 254–260 (doi: 10.1038/nn809).

Voets, T. (2014). TRP channels in thermosensation. In: B. Nilius, V. Flockerzi (eds.) *Mammalian Transient Receptor Potential (TRP) Cation Channels.* Handb. Exp. Pharmacol., Springer, *223*, 729-741 (doi: 10.1007/978-3-319-05161-1_1).

Voets, T., Droogmans, G., Wissenbach, U., Janssens, A., Flockerzi, V. & Nilius, B. (2004). The principle of temperature-dependent gating in cold- and heat-sensitive TRP channels. *Nature*, *430*(7001), 748–754 (doi: 10.1038/nature02732).

Vriens, J., Owsianik, G., Hofmann, T., Philipp, S. E., Stab, J., Chen, X., Benoit, M., Xue, F., Janssens, A., Kerselaers, S., Oberwinkler, J., Vennekens, R., Gudermann, T., Nilius, B. & Voets, T. (2011). TRPM3 is a nociceptor channel involved in the detection of noxious heat. *Neuron*, *70*(30), 482–494 (doi: 10.1016/j.neuron.2011.02.051).

Walters, E. T. (2009). Evolutionary aspects of pain. In: A.I. Basbaum, M.C. Bushnell (eds.) *Science of Pain*, Elsevier, pp. 175-184.

Walters, K. A. & Roberts, M. S. (1993). Veterinary applications of skin penetration enhancers. In: K.A. Walters, J. Hadgraft (eds.) *Pharmaceutical Skin Enhancement.* New York: Marcel Dekker, pp. 345–364.

Waller-Evans, H. & Lloyd-Evans, E. (2015). Regulation of TRPML1 function. *Biochem. Soc. Trans.*, *43*(3), 442-446 (doi: 10.1042/BST 20140311).

Wang, L. X., Niu, C. D., Zhang, Y., Jia, Y. L., Zhang, Y. J., Zhang, Y., Zhang, Y. Q., Gao, C. F. & Wu, S. F. (2019). The NompC channel regulates Nilaparvata lugens proprioception and gentle-touch response. *Insect Biochem. Mol. Biol.*, *106*, 55-63 (doi: 10.1016/j.ibmb. 2018.11.005).

Wang, S., Dai, Y., Kogure, Y., Yamamoto, S., Zhang, W. & Noguchi, K. (2013). Etodolac activates and desensitizes transient receptor potential Ankyrin 1. *J. Neurosci. Res.*, *91*(12), 1591–1598 (doi: 10.1002/jnr.23274).

Wang, W., Zhang, X., Gao, Q. & Xu, H. (2014). TRPML1: An ion channel in the lysosome. In: B. Nilius, V. Flockerzi (eds.) *Mammalian Transient Receptor Potential (TRP) Cation Channels.* Handb. Exp. Pharmacol., Springer, *222*, 631-645 (doi: 10.1007/978-3-642-54215-2_24).

Wasner, G., Naleschinski, D., Binder, A., Schattschneider, J., McLachlan, E. & Baron, R. M. (2008). The effect of menthol on cold allodynia in patients with neuropathic pain. *Pain Med.*, *9*(3), 354–358 (doi: 10.1111/j.1526 -4637. 2007.00290.x).

Wasner, G., Schattschneider, J., Binder, A. & Baron, R. (2004). Topical menthol – as a human model for cold pain by activation and sensitization of C nociceptors. *Brain*, *127*(Pt. 5), 1159–1171 (doi: 10.1093/brain/awh134).

Wechselberger, M., Wright, C. L., Bishop, G. A. & Boulant, J. A. (2006). Ionic channels and conductance-based models for hypothalamic neuronal thermosensitivity. *Am. J. Physiol. Regul. Integr. Comp. Physiol.*, *291*(3), R518–R529 (doi: 10.1152/ajpregu.00039.2006).

Wei, H., Karimaa, M., Korjamo, T., Koivisto, A. & Pertovaara, A. (2012). Transient receptor potential ankyrin 1 ion channel contributes to guarding pain and mechanical hypersensitivity in a rat model of postoperative pain. *Anesthesiol.*, *117*(1), 137–148 (doi: 10.1097/ALN.0b013e31825adb0e).

Wei. H., Koivisto. A., Saarnilehto, M., Chapman, H., Kuokkanen, K., Hao, B., Huang, J. L., Wang, Y. X. & Pertovaara, A. (2011). Spinal transient receptor potential ankyrin 1 channel contributes to central pain hypersensitivity in various pathophysiological conditions in the rat. *Pain*, *152*(3), 582–591 (doi: 10.1016/j.pain.2010.11.031).

Wei, H., Sagalajev, B., Yűzer, M. A., Koivisto, A. & Pertovaara, A. (2015). Regulation of neuropathic pain behavior by amygdaloid

TRPC4/C5 channels. *Neurosci. Letters*, **608**(1), 12-17 (doi: 10.1016/j.neulet.2015.09.033).

Wes, P. D., Chevesich, J., Jeromin, A., Rosenberg, C., Stetten, G. & Montell, C. (1995). TRPC1, a human homolog of a drosophila store-operated channel. *Proc. Natl. Acad. Sci. USA*, **92**(21), 9652–9656 (doi: 10.1073/pnas.92.21. 9652).

Weyer-Menkhoff, I. & Lötsch, J. (2018). Human pharmacological approaches to TRP-ion-channel-based analgesic drug development. *Drug Discov. Today*, **23**(12), 2003-2012 (doi: 10.1016/j.drudis. 2018. 06.020).

White, J. P., Cibelli, M., Urban, L., Nilius, B., McGeown, J. G. & Nagy, I. (2016). TRPV4: Molecular conductor of a diverse orchestra. *Physiol. Rev.*, **96**(3), 911-973 (doi: 10.1152/physrev.00016.2015).

Winkler, P. A., Huang, Y., Sun, W., Du, J. & Lü, W. (2017). Electron cryo-microscopy structure of a human TRPM4 channel. *Nature*, **552**(7684), 200-204 (doi: 10.1038/nature24674).

Wrigley, P. J., Jeong, H. J. & Vaughan, C. W. (2009). Primary afferents with TRPM8 and TRPA1 profiles target distinct populations of rat superficial dorsal horn neurons. *Brit. J. Pharmacol.*, **157**(3), 371–380 (doi: 10. 1111/j.1476-5381.2009.00167.x).

Wu, L. J., Sweet, T. B. & Clapham, D. E. (2010). Current progress in the mammalian TRP ion channel family. *Pharmacol. Rev.*, **62**(3), 381-404 (doi: 10.1124/pr.110.002725).

Wyatt, A., Wartenberg, P., Candlish, M., Krasteva-Christ, G., Flockerzi, V. & Boehm, U. (2017). Genetic strategies to analyze primary TRP channel-expressing cells in mice. *Cell Calcium*, **67**, 91-104 (doi: 10. 1016/j.ceca.2017. 05.009).

Yamaguchi, S. & Muallem, S. (2010). Opening the TRPML gates. *Chem. Biol.*, **17**(3), 209-210 (doi: 10.1016/j.chembiol. 2010.02.009).

Yang, P. & Zhu, M. X. (2014). TRPV3. In: B. Nilius, V. Flockerzi (eds.) *Mammalian Transient Receptor Potential (TRP) Cation Channels*. Handb. Exp. Pharmacol., Springer, **222**, 273-291 (doi: 10.1007/978-3-642-54215-2_11).

Zanotto, K. L., Iodi Carstens, M. & Carstens, E. (2008). Cross-desensitization of responses of rat trigeminal subnucleus caudalis neurons to cinnamaldehyde and menthol. *Neurosci. Letters*, **430**(1), 29–33 (doi: 10.1016/j.neulet.2007.10.008).

Zanotto, K. L., Merrill, A. W., Iodi Carstens, M. & Carstens, E. (2007). Neurons in superficial subnucleus caudalis responsive to oral cooling, menthol, and other irritant stimuli. *J. Neurophysiol.*, **97**(2), 966-978 (doi: 10.1152/jn.00996.2006).

Zeilhofer, H. U. & Brune, K. (2013). Cyclooxygenase inhibitors: Basic aspects. In: S.B. McMahon, M. Kotzenburg, I. Tracey, D.C. Turk (eds.) *Wall and Melzack's Textbook of Pain.* 6th ed., Elsevier, pp. 444-454.

Zhan, L. & Li, J. (2018). The role of TRPV4 in fibrosis. *Gene.*, **642**(1), 1-8 (doi: 10.1016/j.gene.2017.10.067).

Zhang, X., Spinelli, A. M., Masiello, T. & Trebak, M. (2016). Transient receptor potential canonical 7 (TRPC7), a calcium (Ca^{2+}) permeable non-selective cation channel. In: J.A. Rosado (ed.) *Calcium Entry Pathways in Non-excitable Cells.* Chapter 11. Adv. Exp. Med. Biol., Springer, **898**, 251-264 (doi: 10.1007/978-3-319-26974-0_11).

Zhang, X. & Trebak, M. (2014). TRPC7: A diacylglycerol-activated non-selective cation channel. In: B. Nilius, V. Flockerzi (eds.) *Mammalian Transient Receptor Potential (TRP) Cation Channels.* Handb. Exp. Pharmacol., Springer, **222**, 189-204 (doi: 10.1007/978-3-642-54215-2_8).

Zheng, J. (2013). Molecular mechanism of TRP channels. *Compr. Physiol.*, **3**, 221–242 (doi: 10.1002/cphy.c120001).

Zholos, A. V. (2014). TRPC5. In: B. Nilius, V. Flockerzi (eds.) *Mammalian Transient Receptor Potential (TRP) Cation Channels.* Handb. Exp. Pharmacol., Springer, **222**, 129-156 (doi: 10.1007/978-3-642-54215-2_6).

Zhu, X., Chu, P. B., Peyton, M. & Birnbaumer, L. (1995). Molecular cloning of a widely expressed human homologue for the Drosophila trp gene. *FEBS Letters*, **373**(3), 193–198.

Zhu, X., Jiang, M., Peyton, M. J., Boulay, G., Hurst, R., Stefani, E. & Birnbaumer, L. (1996). *Trp*, a novel mammalian gene family essential for agonist-activated capacitative Ca^{2+} entry. *Cell*, **85**(5), 661–671.

Zimmermann, K., Leffler, A., Fischer, M. M., Messlinger, K., Nau, C. & Reeh, P. W. (2005). The TRPV1/2/3 activator 2-aminoethoxydiphenyl borate sensitizes native nociceptive neurons to heat in wild type but not TRPV1 deficient mice. *Neuroscience*, **135**(4), 277–1284 (doi: 10.1016/j.neuroscience.2005.07.018).

Zou, Zh. G., Rios, F. J., Montezano, A. C. & Touyz, R. M. (2019). TRPM7, magnesium, and signaling. *Int. J. Mol. Sci.*, **20**(8), E1877 (doi: 10.3390/ijms20081877).

Zubcevic, L., Herzik, M. A., Wu, M., Borschel, W. F., Hirschi, M., Song, A. S., Lander, G. C. & Lee, S. Y. (2018). Conformational ensemble of the human TRPV3 ion channel. *Nature Commun.*, **9**(1), art. 4773 (doi: 10. 1038/ s41467-018-07117-w).

Zygmunt, P. M. & Högestätt, E. D. (2014). TRPA1. In: B. Nilius, V. Flockerzi (eds.) *Mammalian Transient Receptor Potential (TRP) Cation Channels.* Handb. Exp. Pharmacol., Springer, **222**, 583-630 (doi: 10.1007/978-3-642- 54215-2_23).

INDEX

A

acetylsalicylic acid (ASA), 128, 151, 159
acrolein, 71, 81, 127, 132, 135
action potential, 19, 27, 37, 68, 77
activating peptide (AP), 52, 131, 132
adaptive immunity, 46, 47
addiction, 119
adipocytes, 36
agonists, 1, 20, 23, 25, 44, 46, 53, 56, 58, 66, 80, 81, 82, 85, 86, 91, 92, 97, 110, 115, 116, 117, 121, 122, 123, 126, 127, 129, 132, 133, 135, 143, 159, 164, 170, 176
allicin, 18, 70, 71, 87, 158
allodynia, 80, 81, 82, 85, 86, 87, 89, 91, 93, 101, 105, 109, 110, 127, 135, 153, 155, 164, 171, 173
amygdala, 26, 27
animal behavior, 1, 79, 149
animal models, 22, 23, 30, 114
ankyrin, 3, 10, 18, 19, 20, 31, 49, 52, 81, 168, 173, 174
anterior hypothalamus, 66, 143
antipyrine, 121, 127
anxiety, 27, 114

anxiety disorders, 114
apoptosis, 30, 54, 62
arachidonic acid, 70, 72, 120, 126
arrhythmias, 37
arthralgia, 116
arthropathy, 58
asthma, 29, 114, 119
astrocytes, 58
autophagy, 35, 45, 48
axon guidance, 27

B

B cell lymphoma, 37
backache, 129
behavioral thermoregulation, 43, 67, 68
biogenesis, 45
blood cells, 23, 29
body temperature, 34, 65, 66, 67, 116
bradykinin, 73, 77, 91, 114, 141, 144, 168
brainstem, 43
breast cancer, 30, 62
Brugada syndrome, 37
burning pain, 75, 85, 92

C

calcineurin, 62
calcitonin gen-related peptide (CGRP), 36, 43, 97, 114, 150
calcium responses, 127, 128, 130, 131, 132
calmodulin, 10, 20, 23, 27, 55, 60, 62
camphor, 56, 70, 73, 80, 160
cancer cells, 30, 38, 54, 62
cancer metastasis, 38
cannabinoids, 114, 129, 139, 154
capsaicin, 52, 53, 70, 72, 73, 75, 80, 85, 90, 92, 93, 94, 95, 96, 97, 107, 110, 115, 116, 121, 122, 124, 126, 127, 128, 131, 132, 134, 143, 144, 145, 150, 151, 156, 157, 159, 168, 170
capsazepine, 131, 132, 133, 134
carcinogenesis, 14, 44
carcinoma, 48
cardiac block, 37
cardiac cells, 25
cardiac fibroblasts, 41
cardiomyocytes, 41
cardiovascular system, 2, 22, 25, 32, 58
carrageenan, 127, 129
carvacrol, 18, 56, 70, 71
cation-permeable channels, 1, 14, 45, 79
cell differentiation, 26, 41, 47
cell migration, 22, 27
cerebellum, 26
cerebral ischemia, 29, 42, 168
channelopathies, 57, 114, 137, 163
Charcot-Marie-Tooth disease, 59
chemical irritants, 18, 79
chemotherapeutics, 120
Chinese hamster ovary (CHO) cells, 33, 76, 78
chronic inflammation, 138
cinnamaldehyde, 18, 71, 77, 80, 81, 110, 115, 116, 159, 161, 171, 175

cold pain, 43, 81, 86, 90, 92, 95, 99, 105, 107, 109, 141, 157, 173
cold sensation, 32, 77, 109, 111, 147
cold sensing, 27, 166
colon, 58, 62, 70
constipation, 119
cortex, 42
cough, 53, 59
cryo-electron microscopy (cryo-EM), 9, 11, 13, 21, 52, 55, 137, 153, 158
cyclooxygenase, 119, 126, 153, 175
cyclooxygenase enzymes (COX1 and COX2), 119, 120, 126, 129, 166
cysteine residues, 132
cystic fibrosis, 59
cystitis, 119
cytokines, 14, 34, 36

D

Darier–White skin disease, 23
deafness, 47
dermatitis, 55, 119
dermatoses, 56
desensitization, 27, 108, 129, 141, 143, 145, 147, 151, 166, 168, 175
diabetes, 39, 59, 172
diabetic nephropathy, 23
diclofenac, 121, 122, 123, 125
digestive tract, 38
dipyrone, 121, 127, 161
dorsal root ganglion (DRG), 18
drug development, 21, 113, 115, 117, 137, 143, 146, 166, 174
drug discovery, 11, 23, 115, 126, 160, 170
Duchenne muscular dystrophy, 23
dysmenorrhea, 129
dysplasia, 58

E

edema, 29, 59, 138

effector organs, 65
electrophilic receptor, 18
embryonic development, 13, 41, 148
endometrium, 62
endoplasmic reticulum, 10, 22, 51
endosome-lysosome interaction, 47
endosomes, 45, 46, 47, 48
endothelial permeability, 26
endothelium, 24
environmental irritants, 77, 142
epididymis, 61
epileptogenesis, 26
estrogen, 60
etodolac, 127, 153, 173
exocrine tissues, 61

F

Familial Episodic Pain Syndrome, 114
fear, 27
fibromyalgia, 117
fibrosis, 59, 138, 175
formalin, 71, 81, 120, 127, 129, 161

G

gastrointestinal complications, 120
gastrointestinal contractility, 26
gingerol, 18, 70
glioblastoma, 54
glioma, 54
Glucuronide metabolites, 134
gout, 120, 159

H

hair growth, 55, 56
head injury, 37
headache, 116, 129, 150
hearing, 49
heart failure, 30, 59
heat receptors, 7, 69, 157

heat-sensitive nociceptors, 68
hematocrite, 23
hemoglobin, 23
hemolysis, 24
hepatic fibrosis, 59
herbal remedies, 116
hippocampus, 26, 42
homeostasis, 14, 21, 32, 38, 40, 41, 45, 52, 60, 167, 171
homeotherms, 65
hydrogen peroxide, 71, 132
hyperalgesia, 59, 80, 81, 82, 83, 85, 86, 87, 89, 90, 92, 95, 96, 109, 121, 122, 123, 124, 125, 126, 127, 128, 130, 146, 155, 157, 163, 164, 165, 169, 170, 171
hyperpolarization, 25, 72
hypertension, 58, 59
hyperthermia, 53, 115, 116
hypothalamic preoptic area (POA), 66, 67, 68
hypothalamus, 43, 73, 74
hypotonicity, 57, 58
hypoxia, 42, 58, 168

I

ibuprofen, 129, 131, 132, 134, 135, 146
ibuprofen-acyl glucuronide (IAG), 129, 130, 131, 132, 133, 134, 135
icilin, 71, 98, 131, 132
immune cells, 25, 46, 54, 164
immune system, 23, 34, 77
indomethacin, 131, 132, 135
inflammation, 29, 55, 56, 86, 90, 91, 119, 120, 129, 134, 135, 165, 170
inflammatory pain, 119, 129, 135
inherited cardiac diseases, 37
inherited pain syndrome, 114
innate immunity, 54
insulin, 34, 36, 38, 39
insulin secretion, 34, 38, 39
intracellular vesicles, 22

itch, 3, 55, 56, 81, 114, 139, 164

K

keratinocytes, 55, 56, 57, 73, 74
keratomas, 57
ketorolac, 121, 122, 123, 125, 159
knockout animals, 17, 30
knockout mice, 23, 26, 41, 58, 60, 81, 91, 93, 97, 98, 109

L

leukemia, 35, 48
leukocytes, 41
ligands, 15, 19, 52, 54, 55, 57, 98, 154
limbic structures, 43
lipid sensor, 54, 167
lipofuscinoses, 48
locomotion, 49
long-term potentiation, 21
lornoxicam (xefocam), 121, 122, 123, 125
lung, 18, 22, 29, 39, 40, 41, 48, 59, 62, 70, 77, 147
lung fibrosis, 29
lung metastasis, 39
lysine residues, 132
lysosomal storage disorder (LSD), xii, 45, 48, 82
lysosomes, 45, 46, 47, 48

M

macrophages, 48
mast cells, 25
mechanosensation, 18, 49, 145
melanoma, 33, 35, 158
membrane depolarization, 19, 20, 29, 98
meningeal nociceptors, 43
mental retardation, 26
menthol, 42, 44, 70, 71, 76, 80, 98, 99, 100, 101, 102, 104, 105, 106, 107, 108, 109, 110, 115, 116, 130, 131, 132, 141, 142, 143, 145, 147, 148, 149, 151, 152, 154, 155, 158, 161, 165, 166, 168, 173, 175
microbial pathogens, 77
migraine, 43, 98, 142, 150, 161
mucolipidosis, 44, 45, 47, 51
mucolipins, 46, 48, 156
multiple kidney disease, 114
multiple sclerosis, 37
muscular pain, 129
mustard oil, 18, 80, 81, 87, 97, 115, 154, 159, 167, 168
myocardial fibrosis, 59
myocardium, 30

N

nasal epithelia, 55
neuralgia, 116
neuroblastoma, 35
neuronal cell death, 42
neuropathic pain, 28, 91, 98, 107, 114, 116, 117, 120, 127, 153, 161, 165, 173, 174
neuropathy, 27, 59, 116, 117, 148
neurosecretion, 21
nitric oxide, 25, 26, 71
nociceptive neurons, 35, 97, 99, 107, 168, 169, 176
nociceptive responses, 133, 134
nociceptors, 7, 43, 68, 76, 80, 81, 86, 92, 96, 97, 107, 108, 109, 115, 120, 134, 155, 173
nodose ganglion (NG), 18, 27, 73, 74, 77, 108
non-steroidal anti-inflammatory drugs (NSAIDs), 71, 113, 119, 120, 121, 122, 123, 124, 125, 126, 127, 128, 129, 130, 139, 153, 163, 164, 170
noxious cold sensations, 72
noxious heat, 28, 36, 52, 54, 69, 72, 73, 75, 81, 85, 93, 97, 105, 107, 108, 128, 167, 171, 172

nucleotide polymorphisms, 60

O

olfactory epithelium (MOE), 39, 165
olfactory sensory neurons (OSNs), 39
olfactory system, 38
Olmsted syndrome, 57
opiates, 119
opioids, 113
osmoregulation, 53
osteoarthritis, 120
osteoblasts, 41
ovaries, 62
oxidative stress, 23, 34

P

pain, 1, 2, 3, 14, 27, 34, 44, 49, 52, 53, 55, 56, 59, 68, 72, 75, 79, 80, 81, 85, 86, 90, 91, 92, 93, 95, 98, 107, 109, 111, 113, 114, 115, 116, 117, 119, 120, 121, 129, 134, 135, 138, 139, 142, 143, 144, 145, 146, 148, 149, 150, 152, 153, 154, 155, 157, 159, 160, 161, 162, 163, 164, 165, 166, 168, 169, 170, 172, 173, 174, 175
pain behavior, 28, 53, 59
pain management, 1, 2, 114
pain models, 115, 117
pain pathway, 113, 119, 144
pancreas, 61, 62, 70
pancreatic beta-cells, 36
pancreatic fibrosis, 59
paraneoplastic retinopathy (PR), 34, 163, 174, 175
parathyroid hormone, 40, 60
para-ventricular nucleus, 67
pathological pain, 115
peripheral neuropathy, 27, 59, 153
phosphatidyl-inositol biphosphate (PIP2), 20, 38, 98, 150
phospholipase C (PLC), 19, 26, 30, 78

phospholipases, 119
phospholipids, 120
physical therapy, 120
pituitary gland, 36
placenta, 61
plasma membrane, 15, 18, 21, 22, 34, 35, 46, 47, 51, 54, 61, 120
plasmin, 60
platelets, 41, 48
polycystic kidney disease, 50
postherpetic neuralgia, 53, 116
preoptic nucleus, 66
proliferation, 14, 21, 22, 26, 35, 54, 59, 62
propyphenazone, 121, 127
prostacyclin, 120
prostaglandins, 120, 121, 129
prostaglandins (PGs), 120, 129
prostanoids, 120, 121, 129
prostate, 35, 37, 42, 44, 54, 61, 62, 98, 163, 171
prostate cancer, 37, 42, 54, 171
prostate-specific antigen (PSA), 44
protein kinase C, 30
protein kinases, 27, 30, 39, 97, 119, 129
protein kinases A, 27, 119
pruritus, 57, 119
pulmonary disease, 29, 59
pulmonary fibrosis, 59
pyrazolone derivatives, 121, 127, 161

R

resiniferatoxin (RTX), 52, 53, 70
respiratory airways, 29
respiratory depression, 119
respiratory system, 38, 114
retina, 26
rheumatic disease, 117
rheumatoid arthritis, 120

S

salivary gland, 61
scaffolding proteins, 27
sciatic nerve ligation, 127
second messenger pathway, 78
second messengers, 21
seizures, 23, 27, 30
sensitization, 56, 59, 66, 85, 90, 95, 97, 107, 108, 115, 120, 145, 155, 173
sensory neurons, 18, 19, 26, 34, 42, 43, 52, 53, 68, 69, 73, 85, 91, 98, 99, 147, 151, 153, 166
sensory receptors, 65
sensory transduction, 3, 14, 20, 49, 80, 157
septum, 43, 66
signal transducers, 65
signal transduction, 3, 45
single-nucleotide polymorphism (SNP), 26
skeletal dysplasias, 59
skeletal muscle, 25, 67
smooth muscle, 24, 25, 27, 29, 36, 41, 58
spared nerve injury (SNI), 27
sphingolipidoses, 48
spinal nerve ligation, 120
stroke, 37
sweat gland, 61
synaptic plasticity, 21
systemic administration, 117
systems therapeutics, 113, 146

T

tachyphylaxis, 128, 159
testicule, 62
testosterone, 60, 98
thalamic reticular nucleus, 43
thermal nociception, 73, 97, 152
thermo-genesis, 67
thermoregulation, 53, 66, 67, 68, 74, 143, 145, 161
thermosensation, 1, 7, 8, 52, 66, 74, 143, 145, 160, 161, 172
thermo-TRPs, 7, 52, 56, 65, 69, 70, 80, 81, 110, 115, 117, 170
thromboxane, 120
thymol, 18, 70, 80
tolerance, 119, 139, 170
toothache, 129
topical application, 81, 85, 86, 90, 91, 107, 115, 117, 152, 155
translational medicine, 137
trigeminal ganglion (TG), 18, 73
tubular factors, 60

U

urogenital tract, 43, 98
uromodulin, 60

V

vanillotoxins, 52
vascular tone, 24, 26
vasodilatation, 25, 36, 67
vesicular trafficking, 45
vessel tone, 59
voltage-gated Ca^{2+}-channels, 25
voltage-gated ion channels, 114
vomero-nasal organ (VNO), 23, 39

W

warm-sensitive currents, 66
warmth-sensitive neurons (WSNs), 67, 68
wasabi, 18, 87
wound healing, 55

Related Nova Publications

NAPROXEN: CHEMISTRY, CLINICAL ASPECTS AND EFFECTS

EDITOR: Judson Horner

SERIES: Pain Management – Research and Technology

BOOK DESCRIPTION: In *Naproxen: Chemistry, Clinical Aspects and Effects,* a compilation of the research developed in the past decades on synthetic receptors for naproxen is presented. Naproxen receptors proved their usefulness in chiral separation of the racemate and in other instances of supramolecular chemistry and pharmacy.

SOFTCOVER ISBN: 978-1-53614-129-0
RETAIL PRICE: $82

MANAGEMENT OF POSTOPERATIVE PAIN AFTER BARIATRIC SURGERY

EDITOR: Jaime Ruiz-Tovar, M.D., Ph.D. (Department of Surgery, Universidad Autónoma de Madrid, Spain)

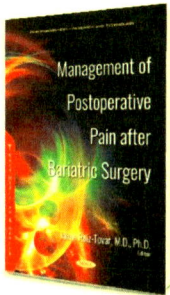

SERIES: Pain Management – Research and Technology

BOOK DESCRIPTION: Patients undergoing bariatric surgery are special subjects, as they present different conditions that make it more difficult to facilitate correct postoperative management. A medical staff is often not used to managing these patients and they do not consider that different measures or doses of drugs should be employed.

HARDCOVER ISBN: 978-1-53614-284-6
RETAIL PRICE: $95

Related Nova Publications

BEHAVIORAL STUDY OF 'NON-OPIOID TOLERANCE'

AUTHORS: Merab G. Tsagareli, Nana Tsiklauri (Ivane Beritashvili Institute of Physiology, Tbilisi, Georgia)

SERIES: Pain and its Origins, Diagnosis and Treatments

BOOK DESCRIPTION: Numerous anatomy-physiological studies have revealed a number of brain structures involved in the shaping of pain and endogenous analgesia. This book presents and examines current research discovered in a behavioral study of 'non-opioid' tolerance.

SOFTCOVER ISBN: 978-1-62100-033-4
RETAIL PRICE: $65

WHY 40%-80% OF CHRONIC PAIN PATIENTS ARE MISDIAGNOSED AND HOW TO CORRECT THAT

AUTHOR: Nelson H. Hendler, MD (Former Assistant Professor of Neurosurgery, Johns Hopkins University School of Medicine, Baltimore, MD, US)

SERIES: Pain Management - Research and Technology

BOOK DESCRIPTION: The book addresses conceptual methods of problem solving, as they are applied to medicine. This book is designed to be the "Freakanomics" for medicine. Many research reports document that 40%-80% (or more, for certain disorders) of chronic pain patients are misdiagnosed. The leading causing of this failure is inadequate history taking, and the use of the wrong medical tests.

HARDCOVER ISBN: 978-1-53612-617-4
RETAIL PRICE: $230